ICME-13 Monographs

Series editor

Gabriele Kaiser, Faculty of Education, Didactics of Mathematics, Universität Hamburg, Hamburg, Germany

Each volume in the series presents state-of-the art research on a particular topic in mathematics education and reflects the international debate as broadly as possible, while also incorporating insights into lesser-known areas of the discussion. Each volume is based on the discussions and presentations during the ICME-13 Congress and includes the best papers from one of the ICME-13 Topical Study Groups or Discussion Groups.

More information about this series at http://www.springer.com/series/15585

Jason Silverman · Veronica Hoyos
Editors

Distance Learning, E-Learning and Blended Learning in Mathematics Education

International Trends in Research and Development

 Springer

Editors
Jason Silverman
School of Education
Drexel University
Philadelphia, PA
USA

Veronica Hoyos
National Pedagogical University
Mexico City, Distrito Federal
Mexico

ISSN 2520-8322   ISSN 2520-8330 (electronic)
ICME-13 Monographs
ISBN 978-3-030-08101-0   ISBN 978-3-319-90790-1   (eBook)
https://doi.org/10.1007/978-3-319-90790-1

Printed on acid-free paper

This Springer imprint is published by the registered company Springer International Publishing AG
part of Springer Nature
The registered company address is: Gewerbestrasse 11, 6330 Cham, Switzerland

Contents

Chapter 1
Research on Technologically Mediated Mathematics Learning at a Distance: An Overview and Introduction

Jason Silverman and Veronica Hoyos

Abstract In this chapter, we provide an overview and introduction to this monograph, which reports on the work of an international group of scholars that joined together at the 13th International Congress on Mathematics Education to share and build on current and emerging research in distance learning, e-learning and blended learning in mathematics. We share work that emerged from *Topic Study Group 44: Distance learning, e-learning, blended learning*, including research and development in the use of digital teaching and learning platforms, usage of this technology to scaffold mathematics instruction and tutoring, novel interfaces for communicating and analyzing student thinking, and specialized mathematics teacher education platforms.

Keywords Research on electronic and distance learning · Teaching and learning platforms · Scaffolding mathematics instruction

This book emerged from the Topic Study Group 44 at the 13th International Congress on Mathematics Education, ICME13, held in Hamburg, Germany on 2016, from July 24th to 31th, where an international group of scholars joined together to share and build on current and emerging research in distance learning, e-learning and blended learning. Specifically, in TSG44 we sought to push on the boundaries of what was known on distance education, e-learning and blended learning and teaching of mathematics through an examination and discussion of recent research and development through these modalities and the common factors that cut across them.

J. Silverman (✉)
Drexel University, Philadelphia, USA
e-mail: silverman@drexel.edu

V. Hoyos
National Pedagogical University, Mexico City, Mexico
e-mail: vhoyosa@upn.mx

The papers published in this monograph are revisions and extensions of the original papers presented during the TSG 44 sessions and reflect additional work carried out by all participant authors after the conference ended. The monograph is organized in four parts: The first part presents two chapters that focus on the incorporation of new technologies into mathematics classrooms through the construction or use of digital teaching and learning platforms (see chapters by Mundt & Hartman, and Hoyos et al., in this book). The second part presents a wide range of perspectives on the study and implementation of different tutoring systems and/or computer assisted math instruction (see Chaps. 4–6 in this book, correspondingly authored by Chekour; Liang et al.; and Landenfeld et al.). The third part presents four new innovations in mathematics learning and/or mathematics teacher education that involve the development of novel interfaces' for communicating mathematical ideas and analysing student thinking and student work (see Chaps. 7–10, authored by Albano & Dello-Iacono; Nakamura et al.; Matranga, Silverman, Klein & Shumar; and Crisan). Finally, the fourth part presents latest work on the construction and implementation of new MOOCs and rich media platforms accomplished to carry out specialized mathematics teacher education (see chapters authored by Avineri et al.; and Chazan et al., in this book).

1.1 Overview of Parts and Chapters

1.1.1 Part I: E-Learning and Blended Learning of Mathematics

Chapter 2 introduces the reader in the construction of the $e{:}t{:}p{:}M^{®}$ platform, developed by Mundt and Hartman from 2012 to 2016, at the University of Education Karlsruhe (Germany). The authors worked to improve the quality of higher education instruction through a platform that integrates digital and internet-based technologies into regular (brick-and-mortar) classes. Specifically, the authors looked at the articulation of online material and other technology to enable a variety of options to be implemented in a blended-course format through the $e{:}t{:}p{:}M^{®}$ approach. Additionally, teachers can modify the content and also student activities can be tracked and reported for a continuous evaluation and improvement.

The potential benefits of this platform are clear as students can find materials for specific content as well as document occasions of interactions with the materials, colleagues, a mentor and/or the teacher; and teachers can create his/her materials specific to their individual needs. In addition to sharing the $e{:}t{:}p{:}M^{®}$ development and model, Mundt and Hartman's chapter provides additional data and analysis regarding the usage of the platform, for example measuring student accesses to the platform as well as understanding relationships between interactions with the content and the archived online lessons.

Chapter 3, by Hoyos et al., from the National Pedagogical University in Mexico City, addresses opportunities and challenges posed by the teaching and learning of mathematics through digital platforms. Specifically, the chapter focuses on the design and implementation of several different mathematics learning environments that provided new teaching and learning opportunities for students in hybrid environments. In this work, the authors establish a relationship between student mathematical attainments and digital tools functionalities, by means of the elaboration of teaching cycles that have influenced the design of the activity and students' learning improvement.

In the chapter by Hoyos et al., the authors documented usage of digital platform tools for the administration, storage and deployment of resources and for facilitating interactions within the resources stored in the platform. One important consideration in this chapter is the teaching challenges of attending to the promotion of reflection processes during the resolution of math problems in an online environment. This paper documents the challenges when teaching and learning of mathematics were completely online and mediated by technology.

1.1.2   Part II: Online Environments and Tutoring Systems for Leveling College Students' Mathematics

In Chap. 4, A. Chekour, from the University of Cincinnati-Blue Ash College in USA, describes an effort to utilize technology to support more effective developmental mathematics learning and teaching. The chapter compares the academic performance of students enrolled in developmental mathematics sections that utilize computer-assisted instruction with those using traditional instruction. Results show the potential of the computer-assisted instruction, both in aggregate and for both males and females separately. The chapter also discusses the challenges and opportunities for incorporating computer assisted instruction into university mathematics classes.

Chapter 5 by Liang et al., from the University of Hong Kong in Hong Kong, presents an evaluation of the different user behaviours on an e-learning platform for students with different levels of calculus knowledge. Having collected data from a sample of 225 students who have used the platform, which includes both video lessons and assessments, the authors focus on student interaction with a supplemental (i.e. not required) e-learning system. Results highlight relationships between activity in the system and student performance on standard examinations. For example, they document that while students with the necessity and urgency to catch up (i.e. with less prior knowledge on calculus) tend to be more active on the e-learning platform in general, many of them tend to ignore the importance the quizzes, which are designed to provide practice and support the development of fluency with the contents at hand.

In Chap. 6, Landenfeld and her colleagues from the Hamburg University of Applied Sciences in Germany discuss the online learning environment via MINT, which was designed to provide differentiated mathematics support to undergraduate science and engineering students. Assessment results allow for the system to recommend differentiated "paths," that can include selections of video tutorials, learning activities and exercises and the "Personal Online Desk" provide a view for the student and others to view progress. In addition to sharing details of the system, this chapter also shares results of analysis of student engagement with the environment.

1.1.3   Part III: Innovations on E-Math Learning and Teaching

Chapter 7, by Albano & Dello-Iacono, from the University of Salerno in Italy, presents an innovative approach to competence-based mathematics learning, through the use of a digital platform named DIST-M (Digital Interactive Storytelling in Mathematics). DIST-M allows learners to define a model where the roles of participants and the sequence of activities promote cognitive, socio-cognitive and metacognitive processes. In this platform, students are engaged in activities within a storytelling experience. The authors used both experiential and discursive approaches to mathematics learning, integrating individual and social tasks, defined by external scripts. The development DIST-M was based on the assumption that such environment can be arranged in a way that a good exploitation of platform tools and a well-structured collaboration among peers can act as an expert support to students in achieving their learning goal. The environment also supports the exploration of specific mathematics content—representation and management of graphics and descriptive statistics, in the case of this paper—in spatial activities that the authors have designed around thematic contexts, such as the discovery of a new planet.

In Chap. 8, Nakamura et al., working in three different universities from Japan (the Nagoya University, the Mukogawa Women's University and the Nihon University) and including the participation of the Sangensha LLC., and the Cybernet Systems Co., Ltd., address two challenges that instructors encounter when implementing the e-learning systems that are prominent in Japan: entry of mathematical symbols and equations and the development of content for specific courses and content. The authors share details of two input interfaces that integrate with commonly used e-learning systems and allow students to input mathematical symbols and equations using both computer and mobile devices, importantly, to address the unique challenges of mathematics e-learning using mobile devices (tablets and mobile phones). These input interfaces are one component of the *MeLQS* e-learning, question specification allows for questions to be cross-platform, and users from many different e-learning systems collaborate and share questions and tasks, thereby making the use of e-learning systems more generative.

Matranga et al. (Chap. 9), from the California State University San Marcos and the Drexel University in USA, describe an online environment designed to support the emergence of a set of professional practices within a group of mathematics teachers. In the chapter, the authors share the design and features of the environment, highlighting the design process through which it was thought to meet specific use cases identified by teachers. The chapter addresses the challenge of scaling teacher professional development through using technology to emulate a boundary encounter between a group of teachers and an existing community of educators with productive pedagogical practices. Their findings show the promise of this approach, specifically noting the emergence of productive pedagogical practices normative in the target community.

Finally in this part, Crisan (Chap. 10), from the UCL Institute of Education, University College London in UK, discusses her work supporting teachers as they explore how digital technology supports students' understanding and learning of mathematics. Video cases that depict actual student engagement with specific mathematics tasks, including audio and video of students synchronized with recordings of their actual work, were specifically developed for this project and participants engagement with these cases—and the student thinking that are depicted in the cases—was analysed. Crisan reports that persistent engagement with these video cases and the other supports provided in the online context show promise for scaffolding teachers as they analyze student work and develop pedagogical solutions based on this analysis. Using a modified version of the Technological Pedagogical Content Knowledge (TPACK) framework, *Research-informed TPACK* (RiTPACK), Crisan presents additional evidence of teacher development resulting from their engagement with the video cases and online course.

1.1.4 Part IV: MOOC and Rich Media Platform for Mathematics Teacher Education

Chapter 11 by Avineri et al., from three universities in USA (the North Carolina School of Science and Mathematics, the North Carolina State University, and the Middle Tennessee State University) and including the Victoria University in Melbourne (Australia), specify design principles for the implementation of MOOCs for professional development of mathematics teachers, based on recent research on this topic. The chapter documents the design efficiency and discusses specific impacts that participants report on changes into their teaching practices. Specifically, some participants addressed changes to their approach to teaching (e.g., increased focus on concepts as opposed to algorithms), others described how their participation supported their refined attention to and understanding of their students' thinking and their own personal improvement in knowledge of mathematics. According to Avineri and colleagues, the research-based design principles that guided the creation of the *MOOC-ED* courses have afforded educators' choice in professional learning, complemented with relevant, job-embedded activities, access to the perspectives of experts, teachers, and students, and a network of educators learning together around a common content area.

Chapter 12, by Chazan et al., from three institutions in USA (the University of Maryland, the University of Michigan, and the Rowland Hall School at Salt Lake City in Utah), describes how the *LessonSketch* platform has been used to implement a larger project between math teacher educators. In particular, these authors use Grossman's pedagogies of practice to explore how teacher educators are representing practice, decomposing it, and providing opportunities for their students to approximate practice through the curricular artefacts that they are creating. Chazan et al., describe a practice-based approach to helping teachers explore the content of mathematics teacher education, and report the novel ways in which a certain online environment (*LessonSketch* in this case) supports new opportunities for teacher candidates to practice the work of teaching. These authors note that professional development experiences created with these platforms not only have pedagogical characteristics and support learning about teaching, but also have curricular characteristics that help shape what it is that teacher candidates should learn.

1.1.5 Purpose of This Monograph

This book addresses issues of collaboration, equity, access and curriculum in the context of learning and teaching mathematics. For example, Mundt and Hartman focus on the population of students entering Universities (Chap. 2) and propose an online platform such as $e{:}t{:}p{:}M^{®}$ to address the challenges brought forth through significant increases in undergraduate populations and associated challenges in instruction and supervision. This is a consistent role posited by authors in this text utilizing existing course management systems and tools, such as *Moodle* and *Blackboard Learn,* as well as other custom designed platforms. While the vast majority of online platforms offer similar features, such as organization of the content, and integration of external software including e-mail, and discussions, the authors noted that $e{:}t{:}p{:}M^{®}$ approach innovates because it could establish or monitor a relationship between the usage of different mobile technology resources with the blended courses it promoted. These results and others presented throughout this volume confirm the existence of new teaching and learning opportunities when working with students in hybrid environments.

With regards to wholly online mathematics learning and teaching, authors in this volume reported the existence of challenges related with the promotion of reflection processes when teachers or students solve mathematics complex tasks while participating in a course at a distance (see Hoyos et al., and Matranga et al., chapters). Using different contexts and approaches, the authors suggest that effective digital collaboration requires attention to individual's (teacher or student) activity and specific supports to accomplish an epistemological change required in order to engage productively and solve such mentioned tasks. These supports can be included in the computational device, learning environment or otherwise be provided by tutorial intervention.

A second broad theme in this volume is the construction and evaluation of mathematics tutoring systems for supporting college students' persistence and success. Such tutoring systems are essential, both given the growth in undergraduate students and continued issues regarding entering freshman's preparation for college level mathematics. While there are various commercial tutoring environments available, the authors in this volume (Chekour; Liang, et al.; and Landenfeld et al.) notice the benefit of custom designed environments to address specific local constraints and share information about their systems as well as suggestions to improve students' use of an e-learning platform.

The third part of this volume addresses a third theme: innovation in e-learning. In this part, the authors discuss new approaches to mathematics learning and mathematics teacher collaboration through the use of Web platforms and communication tools. Albano & Dello-Iacono introduce a general methodology to support an e-learning-based approach to competence-based mathematics learning. These authors designed and implemented certain computer-supported collaboration scripts aimed to foster students' shift from investigating, reasoning and communicating what they have found. Nakamura et al.'s chapter described and displayed a series of interfaces designed to minimize the challenges of mathematical symbols and syntax in e-learning environments. In Matranga et al. chapter, the authors documented that a specifically designed online collaborative environment had the potential to scaffold teachers' legitimate participation in reform-type conversations and activities that were not common for these individuals without the online supports. Finally, Crisan's chapter provides another example on the use of varied multimedia for teacher development resulting from their engagement with video cases and specific online course.

The fourth part of the book addresses a final theme: the use of online rich media platform for teacher education, including the development and implementation of both visualizations of teaching and specially constructed MOOCs for mathematics teacher education. Two of these applications are discussed in Chaps. 11 and 12, and share theoretical and empirical evidence regarding both the effectiveness of the specific design and medium as well as emerging advancements in this area. As an example, Chazan et al. (Chap. 11) use the mathematics education literature on curriculum to suggest that the curriculum creation process that is underway in teacher education, when it happens online, is influenced by the digital nature of technological artifacts.

This book is a scholarly collaboration on the part of professors, developers and researchers in the broad fields of technologically-enhanced mathematics education and serves as an effort to disseminate significant contributions and share international perspectives on this important and timely area. The book provides an overview of the current state-of-the-art research and shares and discusses emerging work, including trends, ideas, methodologies, and results and represents a special call to continue research and development and to grow a canon of research foundations for distance learning, e-learning and blended learning in mathematics education.

Part I
E-Learning and Blended Learning of Mathematics

Chapter 2
The Blended Learning Concept e:t:p:M@Math: Practical Insights and Research Findings

Fabian Mundt and Mutfried Hartmann

Abstract The chapter outlines the key ideas of the blended learning concept e:t:p:M® and its further development in the field of Higher Mathematical Education. e:t:p:M@Math aims to integrate digital technologies and face-to-face interactions to simultaneously allow personalized and high-quality learning. Both practical teaching experiences as well as research findings will be discussed. One focus is on the description of the self-developed and designed e-Learning environment, its possibilities and further development. Another focus is on the reflection of the practical implementation into everyday teaching, especially the integration with face-to-face seminars. In addition, first research insights will be presented and explained.

Keywords Blended learning · E-learning · Distance learning · Learning analytics

2.1 Introduction: Challenging Trends in Higher Education

In the winter term 2014/2015, the Federal Bureau for Statistics of Germany counted 2.7 million university students—a milestone in the history of the Federal Republic of Germany (SB, 2015). Given that only ten years ago there were far less than 2 million students (Bildungsbericht, 2014), the magnitude of this increase becomes even more significant. Many universities have adopted "bulk-instruction" with heterogeneous student groups and an unfavorable student-to-instructor ratio (Himpsl, 2014). In particular, high-demand introductory courses suffer under these problematic circumstances. Therefore, the quality of education is lacking and the need for reforms is apparent (Asdonk, Kuhnen, & Bornkessel, 2013).

F. Mundt (✉) · M. Hartmann
University of Education Karlsruhe, Bismarckstraße 10, 76133 Karlsruhe, Germany
e-mail: fabian.mundt@ph-karlsruhe.de

M. Hartmann
e-mail: mutfried.hartmann@ph-kalrsruhe.de

© Springer International Publishing AG, part of Springer Nature 2018
J. Silverman and V. Hoyos (eds.), *Distance Learning, E-Learning and Blended Learning in Mathematics Education*, ICME-13 Monographs,
https://doi.org/10.1007/978-3-319-90790-1_2

Since this situation is unlikely to change in the foreseeable future—neither nationally nor internationally (Dräger, Friedrich, & Müller-Eiselt, 2014; Maslen, 2012)—innovative teaching and learning concepts are necessary. In contrast to widely-discussed MOOCs, one very promising approach involves integrating regular class sessions with the opportunities of digital (Internet-) technologies (see Carr, 2012). One specific model that specifically aims at the integration of both class sessions and digital content is e:t:p:M®.

2.2 The Blended Learning Concept e:t:p:M®

e:t:p:M®[1] was developed as an introductory course in education in the winter term 2012 at the University of Education in Karlsruhe by Timo Hoyer and Fabian Mundt. Detailed information about the project and its theoretical framework can be found in Hoyer and Mundt (2014, 2016). The acronym, which indicates the individual parts of the project, are described in depth below.

2.2.1 "e" for E-Learning

The core of the e-learning content consists of 11 studio recorded *online lessons* that have been post-produced according to a creative framework. The lessons are all between 20 and 30 min long and are comprised of a speaker as well as info boards, images, animations and quotations. Additionally, the lessons are structured through so called "Fähnchen" (thematic headlines). The students can access the content via an especially for the e:t:p:M® project developed *responsive web-app* (Fig. 2.1).[2]

Personal annotations can be added to every "Fähnchen" and then downloaded as a PDF-file (Fig. 2.2). Furthermore, the web-app grants access to additional materials (texts, exercises etc.) and does not only contain general information about the class but also an extensive FAQ-area and the possibility to get in touch with the lecturers directly. The web-app also provides the user with a differentiated module for analysis that enables the teacher to track the students' interactions.[3]

---

[1]Project website: http://etpm.ph-karlsruhe.de/demo/ [13.12.2016].

[2]The web-app was developed with the open-source frameworks Laravel, Vue.js, Semantic UI and Video.js.

[3]As a tool for analyzing the interaction, an adjusted version of the open analytics platform "Piwik" is used. All collected data is anonymized. The tracking function can be deactivated from inside the web-app, which is highlighted for the users.

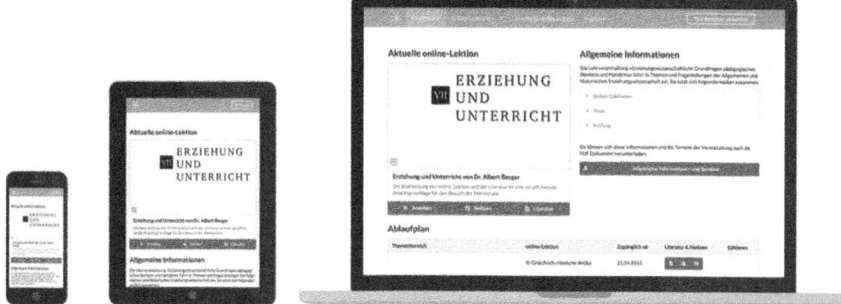

Fig. 2.1 The responsive web-app (original version)

Fig. 2.2 The annotation function of the web-app

2.2.2   *"t" for Text and Theory Based*

Alongside every online lesson, the students are provided with a text that deepens the content (primary as well as secondary literature). In addition to suggested approaches to the text, the file contains questions that will be dealt with during the attended seminar. All texts are formatted uniformly and have been edited for the use in a seminar.

2.2.3   *"p" for Practice-Oriented (and Attendance-Oriented)*

The e-learning content of e:t:p:M® aims at a high personalization of the learning content as well as its integration into the seminars. The latter are comprised of information sessions (lecturers), FAQ-sessions (lecturers) and weekly mentoring sessions (student mentors) (Fig. 2.3).

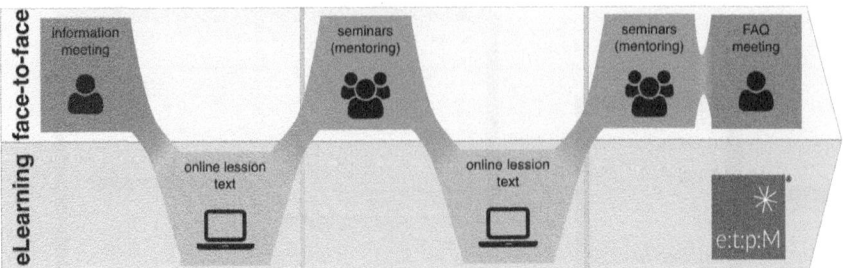

Fig. 2.3 The e:t:p:M® concept

2.2.4 *"M" for Mentoring*

Especially in the beginning of studies at university, the support and care for beginners is of high importance. In addition to subject-specific competences, the students need to acquire a sense to navigate the foreign academic world. In e:t:p:M® the class is separated into smaller groups who will be mentored by a tandem of older students during the semester. The mentors are trained in a specifically designed workshop and receive a certificate after completion.

The program received an award for extraordinary teaching methods in 2013 and was evaluated positively several times.[4]

2.3   Using e:t:p:M® in an Introductory Course in Mathematics

Based on the previously explained challenges for teaching at university and the very positive feedback towards the project e:t:p:M®, the concept is being adapted for other subjects outside the realm of pedagogy. At the moment, the authors work on applying the program to an "Introduction in Mathematics" course, which started in the winter term 2015 (Mundt & Hartmann, 2015). The current evolution of the concept is presented below. Since the contents are more historical and theory oriented, the application of e:t:p:M@Math requires adjustments. The online lessons and web-app, in particular, are being revised extensively to meet the requirements of mathematical learning.

The adaptation is informed and guided by "design-based research methodology" (Wang & Hannafin, 2005). Specifically, it is situated in a real educational context (mathematics introduction) and is focusing on the design and testing of significant interventions (e:t:p:M@Math concept) (see Anderson & Shattuck, 2012). As part of the design process, we refer to contemporary findings in the field of Higher

---

[4]http://etpm-dev.ph-karlsruhe.de/etpm-evaluation/ [13.10.2015].

Education eDidactics (Ertl, 2010; Kerres, 2013) with a special focus on mathematical learning in digitally supported environments (Aldon, Hitt, & Bazzini, 2017; Juan, 2011) and "User Experience Design" (Meyer & Wachter-Boettcher, 2016; Walter, 2011). In this text, we are concerned with the extensions of the web-app.[5] For this reason, we also include a review of existing blended-learning specific tools.

2.3.1 Existing Tools and e:t:p:M@Math Web-App

A review of the current literature and software shows that there are many blended learning concepts in the field of Higher Education, but only few explicit tools. Besides well-established Learning Management Systems like *Moodle*, *OpenOLAT* or *ILIAS* there are some more recent MOOC related platforms like *edX*. A more detailed overview of these and similar resources can be found in Spring et al. (2016) and Ma'arop and Embi (2016). All these solutions offer functionality for blended learning scenarios. Often, these tools require special plugins or add-ons (Kumari, 2016). They also often lack both a good user experience design and context-specific needs (e.g. for mathematics teaching), which goes hand in hand with the overwhelming functionality of the software (Persike & Friedrich, 2016). Hence, it is no surprise that there are also a variety of special and often well-designed tools in addition to the all-embracing systems. These range for example from applications which enable the creation of interactive videos (H5P[6]), deliver the opportunity to brainstorm online (MindMeister[7]) or create entire learning lessons easily (TES Teach[8]).

In contrast to this situation, the e:t:p:M@Math web-app is a blended learning-specific software. This means it integrates modern technologies and ideas, e.g. creating rich interactive video content, with the pedagogical aspects of the e:t:p:M® concept and context specific needs in mind. One example might be personalized annotations optimized for seminar use (see Fig. 2.2) or instant exercise feedback for teachers as outlined below. The web-app can be seen as a continuously developing framework in the sense of the design-based research, where interventions are repeatedly added, evaluated and improved. The web-app itself can also be seen as a research tool.

---

[5]The web-app is developed in the sense of "Agile Software Development" (Dingsøyr, Dybå, & Moe, 2010), which fits perfectly with the design-based research methodology.

[6]https://www.h5p.org [10.7.2017].

[7]https://www.mindmeister.com [10.7.2017].

[8]https://www.tes.com/lessons [10.7.2017].

2.3.2   Research Questions

The review of the current literature and software showed that there is a lack of a well-designed blended learning specific software that integrates modern technologies in a tight didactical way and is also open for further research-based development. Out of this, our focus is on the following overarching question: *How can modern technology enabled options be implemented in the mathematical adaptation of e:t:p:M®?* The following sub-foci organize our discussion of our broad research focus:

(a) How can *interactive content* be integrated in the web-app?
(b) How can *discussions*—a key element of teaching and learning mathematics—be integrated in the web-app?
(c) How can *exercises and tests* be implemented?
(d) How can the teachers *modify and generate content*?
(e) How can student activities be tracked and reported for *continuous evaluation and improvement*?

2.3.3   Series of Interactive Content (Sub-focus a)

An online lesson is not only comprised of just a single video, but contains a series of shorter videos and interactive learning applications. This series of interactive content enables a more differentiated structure of the more abstract, mathematical learning contents. The interactions make it possible for the user to comprehend complex correlations on their own. Current versions of the video environment are shown in Figs. 2.4 and 2.5.

At present, we are considering about at least three different content types:

- Interactive videos
- Exploration exercises
- Test exercises

As you can see in Fig. 2.4 the video environment integrates these new ideas in the existing application. In addition, the concept of "Fähnchen" can be used in both the shorter videos and in exercises. In case of the exercises, the concept has to be adjusted, particularly through structuring each interactive exercise around several tasks. Each of these tasks can be visually and functionally highlighted by one "Fähnchen". Thereby individual notes can be taken while solving the tasks.

To implement interactive exploration and test exercises the open-source software CindyJS[9] is used. CindyJS is a JavaScript implementation of the well-known interactive geometry software Cinderella (Richter-Gebert & Kortenkamp, 2012).

---

[9]http://cindyjs.org [13.12.2016].

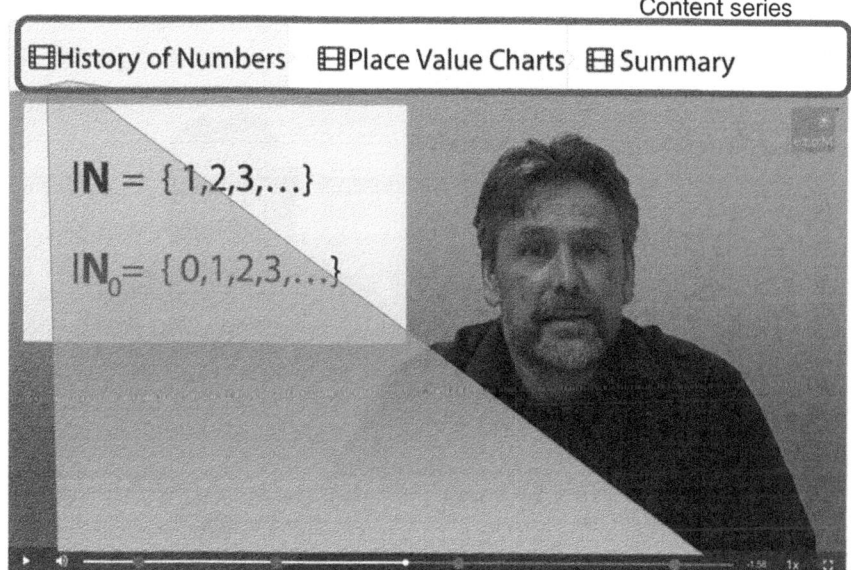

Fig. 2.4 Implementation of content series

The software is mainly used to connect with existing content. However, a modern JavaScript-framework also makes is possible to optimize the learning environment for modern devices, which is very important in the mobile world. Furthermore, it is possible to develop an intuitive content editor, which allows for quick and easy editing of the exercises (Sect. 2.3.4).

2.3.4 Discussions (Sub-focus b)

Especially in mathematics, controversial discussions facilitate learning and create a lot of questions. Hence, the annotations and the series of interactive content are supplemented by discussions. In the discussion forums, which are based on traditional online discussions, students can post questions and answers tailored to the corresponding "Fähnchen". Additionally, the students can 'like' the posts, and lecturers can highlight relevant questions or interesting postings. To keep matters clear, irrelevant postings will be faded out after a while.

Since 2016, we have been testing several forms of in-app discussions. Following the iterative implementation and evaluative paradigm of the design-based research approach we present here both, a first draft (Fig. 2.5) and the current version

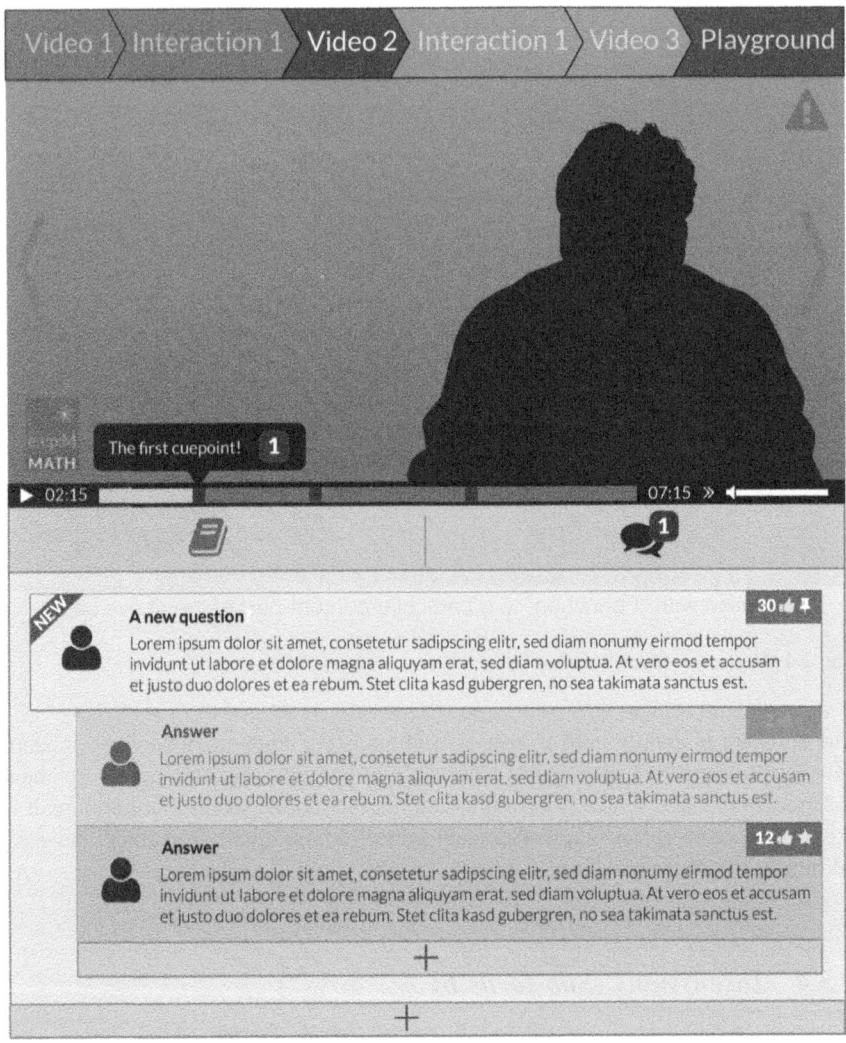

Fig. 2.5 First draft of "discussions"-extension

(Fig. 2.6). As you can see in Fig. 2.6 we are using the popular disqus™ service at the moment. By using an existing service, we gain valuable insights for our own implementation. In addition to technical challenges, it seems important to develop a discussion format which integrates very well with the existing parts of the blended learning concept, particularly the interlink to the face-to-face seminars.

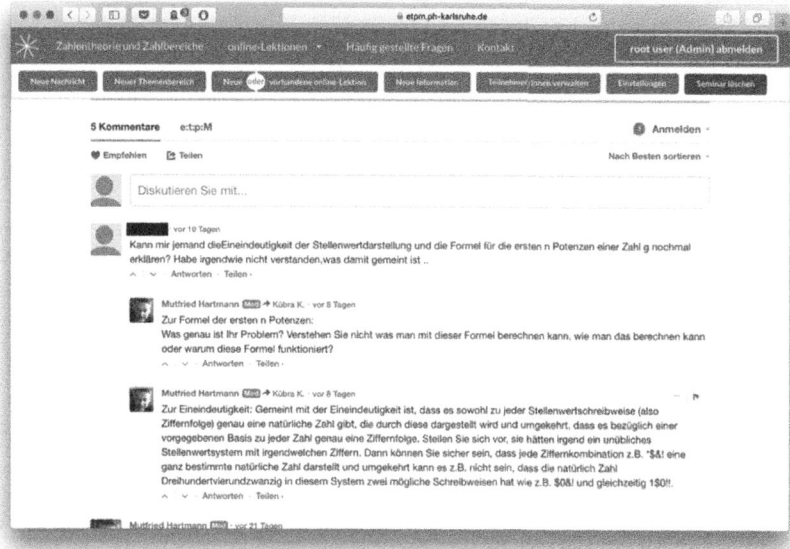

Fig. 2.6 Implementation of discussions (native web-app screenshot, see Fig. 2.8 for translated interface)

2.3.5   Playground (Sub-focus c)

Every online-lesson contains a summarizing event called "playground". The playground consists of practice final exercises and interactions that can be scored in terms of understandability. Through that, a feedback loop is created that enables a focused learning process. Furthermore, a dynamic script can be generated. If desired, the personal annotations and discussions from the forum can be included at the respective place in the script, before it can be downloaded as a PDF (Fig. 2.7).

2.3.6   Content Editor (Sub-focus d)

In addition to functions that deal with teaching content, the easy creation and revising of online materials is an important point. Especially in the context of mathematical settings, where abstract ideas and their representation are focused, content production can become very intensive.

For this reason, we are working to extend the web app with an "editor mode", which makes it possible to create rich learning content quickly and easily. On the one hand, this includes the possibility to add "Fähnchen" and interactive areas (links, graphs, pictures etc.) to video content. An elaborate postproduction is

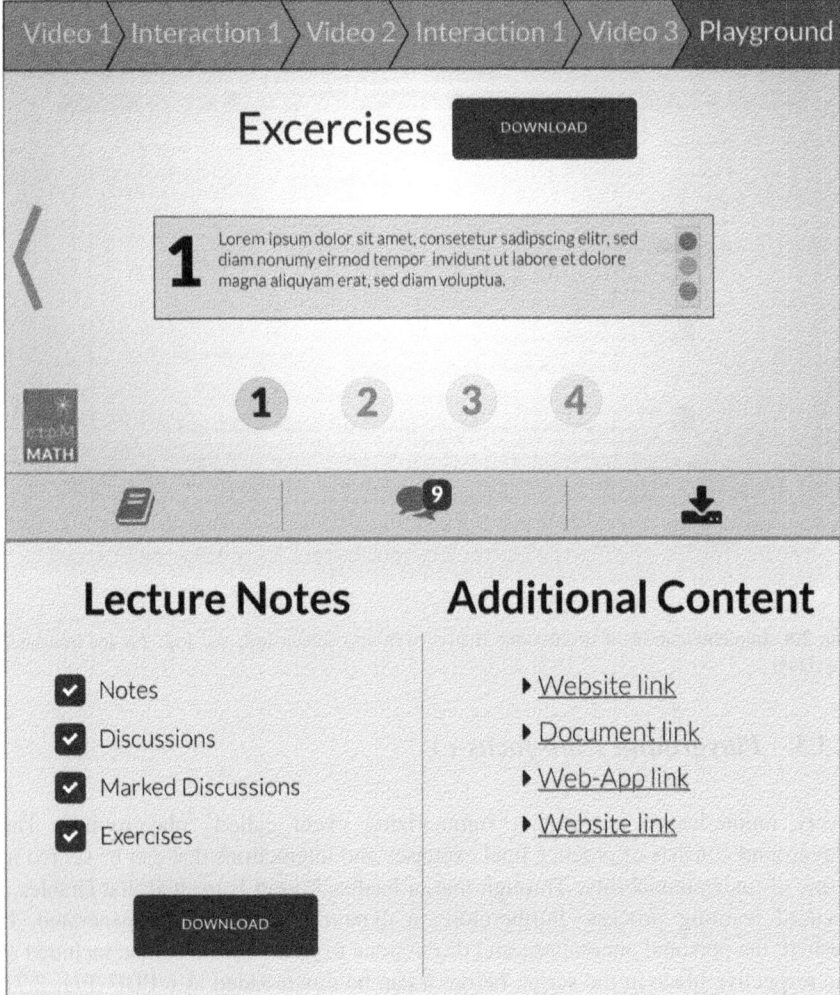

Fig. 2.7 First draft of the "playground"-extension

thereby reduced to the essential and the video content can be changed more easily. A great advantage is also that the teachers can do this by themselves. On the other hand, the interactive applications (exploration and test exercises) will be editable as well.

In this way, the web-app offers teachers the opportunity to create a comprehensive and modern learning environment, which reflects both the requirements of the mathematical subject and those of the learners.

Figure 2.8 shows the current development of the content editor.[10] We are trying to integrate the editor features seamlessly within the user interface. Teal-colored buttons highlight interaction possibilities, and the main features are bundled in the secondary menu bar. Following the agile development paradigm, we are implementing extensions continuously. At the moment we have implemented the following "main features":

- news message system
- FAQ system
- seminar management
- rudimentary online lesson management
- discussion system
- user management
- user role system
- integration in the university ecosystem through LDAP support
- integration of the open analytics platform Piwik (see footnote 3)

In other words, teachers are able to create and manage their own seminars. This includes the possibility to arrange existing online lessons or add new ones, manage seminar participants, upload documents, write news messages and manage discussions through a disqus™ service integration. Further, they can receive statistically prepared live analytics and configure the system setup.

Unfortunately, it is at the moment not possible to create new high-level online lessons without technical knowledge. The next step will be to develop the previously explained online lesson editor based on the already mentioned CindyJS and H5P frameworks.

2.3.7 Continuous Evaluation Strategy (Sub-focus e)

The series of interactive content, the discussions, the playground and the content editor mark the current state of the further development of the e-learning content of e:t:p:M@Math. At the moment, the program—in particular, the discussions and their integration with face-to-face seminars—is being implemented and evaluated. In addition to continuous web-based interaction analysis, we have planned an extensive survey in 2017, when more parts of the introduction course are implemented according to the blended-learning concept. Because we can relate both tools of our analytical framework—which is called "Blended Evaluation" (Mundt & Hoyer, 2017)—we expect to identify detailed learning profiles. In fact, we hope to obtain a solid empirical basis for further design-based decisions. In the following sections, we share some initial results, which focus the web-app interactions.

---

[10]The web-app project, called "Synthesise", is published under the MIT Open Source license. The code is available on GitHub: https://github.com/inventionate/Synthesise. Participation is welcome.

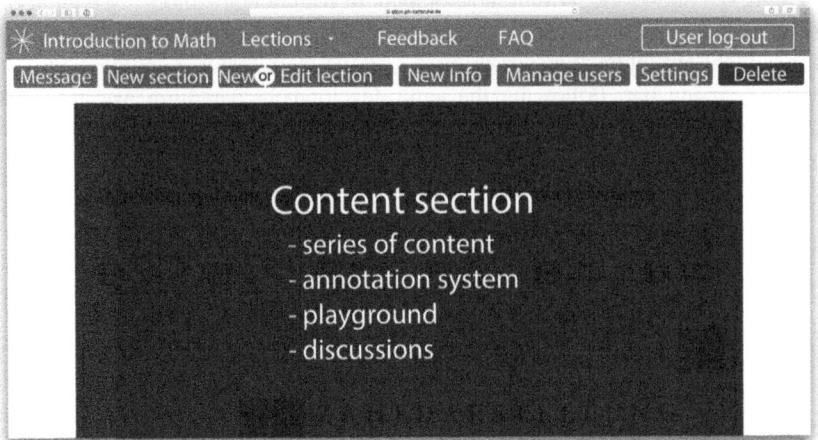

Fig. 2.8 First implementations of the content editor (web-app screenshot, translated)

2.4 First Insights of the Use in Winter Semester 2015/2016

The first three sessions of an "Introduction in Mathematics" course (October 2015 to November 2015) were organized in the e:t:p:M® format. Altogether, there were 168 students who attended the course. In this section, we will present some statistical analysis of the web-app interactions using *Learning Analytics*, which is described as "an educational application of web analytics aimed at learner profiling, a process of gathering and analyzing details of individual student interactions in online learning activities" (NMC, 2016). We are focusing on the web-interactions for two reasons. First, the introduction course is at this time only partly adapted. A comparative analysis, which allows conclusions on the student performance is therefore very difficult. As already said, this is methodically prepared as well as planned for future semesters (see Mundt & Hoyer, 2017). Second, the evaluation of web-interactions (Learning Analytics) provides important information for our agile development and design-based research process.

2.4.1 Evaluation Strategy and Questions

We begin by interpreting some core web-analytics data provided by our Piwik (see footnote 3) installation. Based on these singular insights we apply a more elaborate multivariate approach—the so called Multiple Correspondence Analysis (see Sect. 2.4.3 for additional details and references)—to visualize entangled relationships. In the sense of our design-based research methodology, we seek to identify indicators of particularly effective interventions and developments. In this context, two main questions are addressed:

a. How do students use the web-content?
b. Can web-interactions be condensed into interaction-profiles?

2.4.2 Usage of Web-Content (a)

Figure 2.9 shows the visitors of the web-app form October to November. After a period of initial curiosity, the tool has been used continuously. In particular, one day before the weekly seminars (mentoring), the online lessons were intensively watched.

More important than the number of visitors over time are their content inter-actions (clicks, movements, typed characters). For example, how long and how often videos were played seems to make a difference. In our case, the average playing time was more than 30 min per online lesson and they were often played multiple times (3–4 times). These data may indicate that there was much 'done' with the videos—and hopefully much learned.

Additional data also supports the conjecture regarding active student engage-ment. Overall, the online lessons were played more than 45,000 times, almost as often as they were paused. Nearly 35,000 times the students skipped parts of the video by using some "Fähnchen". At least 14,500 annotations were made by most of the students.

Apart from traffic and interaction data there is some other interesting informa-tion. Looking at the above-mentioned challenge of digitization, it is appropriate to analyze the devices which have been used by the students. In the case of "Introduction in Mathematics", 80% used 'traditional' desktop or laptop computers to access the web-app. 20% used 'modern' smart devices—10% used smartphones and 10% tablets.

Overall, it can be stated that the students have used the web content extensively. Even if no information about the performance of the students could be integrated, the amount of web-interactions, the time spent and the continuity of usage leads to meaningful insights for further research.

However, research methods which focus on the relationship between those particular aspects of content usage are even more useful. They enable the recon-struction of interaction-profiles, which allows well-informed didactical interven-tions as they are intended by the referenced design-based research methodology.

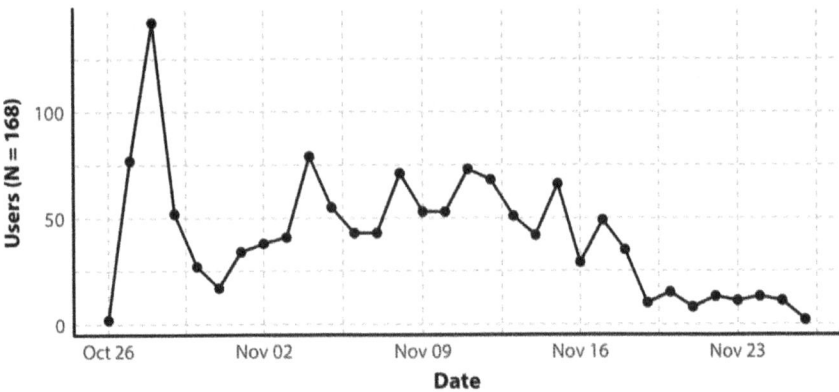

Fig. 2.9 Web-app users (October 2015 to November 2015)

2.4.3 Reconstruction of Interaction-Profiles (b)

In order to reconstruct rich interaction-profiles, we applied a *Multiple Correspondence Analysis (MCA)*. This is a multivariate statistical method, which belongs to a framework known as "Geometric Data Analysis" (Le Roux & Rouanet, 2010). In contrast to popular methods, e.g. Factor Analysis, geometrical methods are exploratory oriented, which is concerned with the construction of "social spaces". That is, an individual level is observed throughout the whole geometrical modeling process (see ibid). Especially in the case of a "Blended Evaluation," this property is of great advantage. It allows both the analysis of the position of each student as well as the reconstruction of group profiles (see Mundt & Hoyer, 2017). Within this chapter, we can only provide brief insights.

2.4.3.1 Dataset and Analysis Toolkit

The analyzed data refer partially to the previously mentioned values. Overall, the correlations between 12 interaction-variables are evaluated. These can be grouped under three headings:

1. Duration of web-app interactions overall (2 variables)
2. Amount of web-app interactions overall without video (5 variables)
3. Amount of video interactions in particular (5 variables)

All variables were categorized in order to be able to carry out an analysis which is as meaningful as possible. The two temporal variables (first heading) were divided into four categories: *very short* (0–7 h), *short* (8–13 h), *long* (14–23 h), *very long* (24–60 h). Likewise, the five web-app counting variables (second heading) as well as the five video counting variables (third heading) were divided in four categories. Web-app interactions are, as already mentioned above, mouse clicks, mouse movements and keyboard input. They were categorized in: *very few* (0–800 interactions), *few* (801–2000 interactions), *many* (2001–4000 interactions), *very many* (4001–20000). Video interactions (playback, pause, jump, speed change) were categorized in: *very few* (0–600 interactions), *few* (601–1800 interactions), *many* (1801–3300 interactions), *very many* (3001–9700 interactions). In total, the dataset consists of 12 columns (one per variable), 168 rows (one per student) and 48 categories (12 variables multiplied by 4 categories).

A standard MCA was performed on this dataset using the free statistical environment R (R Core Team, 2017). To be more specific, the Geometric Data Analysis related packages "FactoMineR" (Lê, Josse, & Husson, 2008) and "factoextra" (Kassambara & Mundt, 2017) were used. In the sense of open and reproducible science, all analysis scripts are available online.[11]

2.4.3.2   MCA Results and Interpretation

Figure 2.10 visualizes the MCA results as a biplot. Both the categorical locations as well as the corresponding positions of the students (grey points) are recognizable. This two-dimensional MCA solution clarifies 86.6% variance, which underlines the significance of the analysis. Usually, solutions over 70% are considered sufficiently meaningful (see Le Roux & Rouanet, 2010). In order to obtain the greatest possible overview, only the barycenters of the variable headings are mapped. For example, the position of "Amount of video interactions: many" results from the positions of the corresponding categories of the five variables of this heading and so on.

Looking at Fig. 2.10 it is striking that the location of the four categorical barycenters of all three headings are always arranged approximately the same. By connecting the points, a kind of parabola becomes visible. As a matter of fact, this effect is a well-known methodological artefact. It reflects the ordinal structure of the variables under investigation (Le Roux & Rouanet, 2010).

In terms of content, the so called "horseshoe-effect" (ibid.) makes clear that students who have a lot of interaction with the web-app and videos also have visited the app significantly longer than the other students. The respective categories are distributed ascending from left to right.

Reflecting the corresponding positions of the students (grey points) four more or less separated interaction-profiles can be identified. These profiles are linked to the temporal intensity and frequency of web-app interactions. Hence, it is almost impossible that people work considerably more with the video elements, but interact rather less with other elements of the web-app. Furthermore, it is noticeable that in particular the less-intense-user-profile is separated sharply (down left). This refers to approximately 25% of the students who have used the web content rarely, which can be interpreted as an indication of a skeptical attitude towards the concept. In contrast, there is a less clearly separated spread of approximately 75% positions

---

[11]https://github.com/inventionate/learning-analytics [11.07.2016]; especially file "mca_mathe.R".

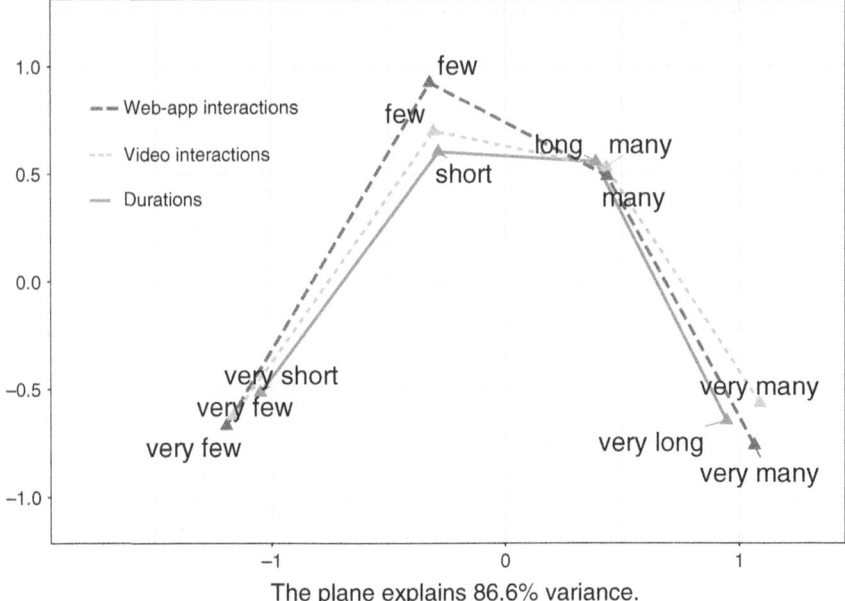

The plane explains 86.6% variance.

Fig. 2.10 MCA biplot on web-app interactions

along the other three categories (short/few to very long/very many). We have begun to outline three additional interaction-profiles, which can be attributed to a positive attitude towards the concept, and are related to a gradual distribution of the inter-action intensity. These three interaction profiles appear to be fluid and permeable. Additional research is underway to further analyze the interaction profiles and their impact.

On the one hand, this analysis confirms the altogether intensive use of the web-app, which has been already discussed in the previous section. On the other hand, a more detailed description and understanding of this initial observation became possible through the Geometric Data Analysis.

2.5   Conclusion

Based on the results presented above, we argue that the high adoption progress clearly signifies the potential of the online system. Particularly, it indicates the potential of the mathematical adaptation of the established blended learning soft-ware. In combination with the previously gained experiences with the original e:t:p: M® concept and the ongoing optimization, we are very optimistic to be able to offer an outstanding mathematical learning environment, and especially one that allows for interesting empirical insights in mathematical learning processes. Our work continues, with currently research focusing on further analysis and evaluation, particularly of successful mathematical learning and its relation to the interaction profiles. Results of this work is forthcoming.

References

Aldon, G., Hitt, F., Bazzini, L., & Gellert, U. (Eds.). (2017). *Mathematics and technology*. Cham: Springer.

Anderson, T., & Shattuck, J. (2012). Design-based research: A decade of progress in education research? *Educational Researcher, 41*(1), 16–25.

Asdonk, J., Kuhnen, S. U., & Bornkessel, P. (Eds.). (2013). *Von der Schule zur Hochschule*. Münster: Waxmann.

Bildungsbericht. (2014). *Bildung in Deutschland 2014*. Bielefeld: wbv.

Carr, N. (2012). The crisis in higher education. *MIT Technology Review Magazine, 115*(6), 32.

Dingsøyr, T., Dybå, T., & Moe, N. B. (Eds.). (2010). *Agile software development*. Berlin/ Heidelberg: Springer.

Dräger, J., Friedrich, J.-D., & Müller-Eiselt, R. (2014). *Digital wird normal*. Gütersloh: CHE.

Ertl, B. (Ed.). (2010). *Technologies and practices for constructing knowledge in online environments: Advancements in learning*. Hershey: IGI Global.

Himpsl, F. (2014). *Betreuer, dringend gefragt*. Die Zeit. 09/2014.

Hoyer, T., & Mundt, F. (2014). e:t:p:M – ein Blended-Learning-Konzept für Großveranstaltungen. In K. Rummler (Ed.), *Lernräume gestalten – Bildungskontexte vielfältig denken* (pp. 249–259). Münster: Waxmann.

Hoyer, T., & Mundt, F. (2016). Den Studienanfang pädagogisch gestalten. Das Blended Learning Konzept e:t:p:M®. In R. Bolle & W. Halbeis (Eds.), *Zur Didaktik der Pädagogik*. Herbartstudien Bd. 6.

Juan, A. (2011). *Teaching mathematics online: Emergent technologies and methodologies*. Hershey: IGI Global.

Kassambara, A., & Mundt, F. (2017). *Factoextra—Extract and visualize the results of multivariate data analyses*. CRAN.

Kerres, M. (2013). *Mediendidaktik* (4th ed.). Berlin: Oldenbourg.

Kumari, S. (2016). Personalised, flexible and blended learning features of moodle-LMS. *Educational Quest- An International Journal of Education and Applied Social Sciences, 7*(1), 53.

Lê, S., Josse, J., & Husson, F. (2008). FactoMineR: An R package for multivariate data analysis. *Journal of Statistical Software, 25*(1), 1.

Le Roux, B., & Rouanet, H. (2010). *Multiple correspondence analysis*. London: SAGE.

Ma'arop, A. H., & Embi, M. A. (2016). Implementation of blended learning in higher learning Institutions: A review of literature. *International Education Studies, 9*(3), 41.

Maslen, G. (2012). Worldwide student numbers forecast to double by 2025. *University World News, 209*. Retrieved 10/2015, from http://www.universityworldnews.com/article.php?story=20120216105739999.

Meyer, E. A., & Wachter-Boettcher, S. (2016). *Design for real life*. New York: ABA.

Mundt, F., & Hartmann, M. (2015). Klasse trotz Masse am Studienanfang – das Blended Learning Konzept e:t:p:M@Math. In H. Linneweber-Lammerskitten (Ed.), *Beiträge zum Mathematikunterricht 2015*. WTM: Münster.

Mundt, F., & Hoyer, T. (2017). Blended Evaluation in der digital gestützten Lehre. In P. Pohlenz (Ed.), *Digitalisierung der Hochschullehre – Hochschullehre in der digitalen Welt: Neue Anforderungen an die Evaluation?* Münster: Waxmann.

NMC. (2016). *Horizon report 2016*. Higher Education Edition. Retrieved 03/2016, from https://library.educause.edu/~/media/files/library/2016/2/hr2016.pdf.

Persike, M., & Friedrich, J.-D. (2016). Lernen mit digitalen Medien aus Studierendenperspektive. *hochschulforum digitalisierung*. 17.

R Core Team. (2017). *R: A language and environment for statistical computing*. Vienna, Austria: RC Team.

Richter-Gebert, J., & Kortenkamp, U. (2012). *The Cinderella.2 Manual. Working with the interactive geometry software*. Berlin/Heidelberg: Springer.

SB. (2015). *Zahlen & Fakten: Studierende insgesamt nach Hochschularten*. Retrieved 10/2015, from   https://www.destatis.de/DE/ZahlenFakten/GesellschaftStaat/BildungForschungKultur/Hochschulen/Tabellen/StudierendeInsgesamtHochschulart.html;jsessionid=EC79F16337CCA9999EB90629995D1E85.cae4.

Spring, K. J., Graham, C. R., & Hadlock, C. A. (2016). The current landscape of international blended learning. *International Journal of Technology Enhanced Learning, 8*(1), 84.

Walter, A. (2011). *Designing for emotion*. New York: ABA.

Wang, F., & Hannafin, M. J. (2005). Design-based research and technology-enhanced learning environments. In *Educational technology research and development* (Vol. 53, No. 4, pp. 5–23). The Netherlands: Kluwer Academic Publishers.

Chapter 3
Challenges and Opportunities in Distance and Hybrid Environments for Technology-Mediated Mathematics Teaching and Learning

Veronica Hoyos, Maria E. Navarro, Victor J. Raggi and Guadalupe Rodriguez

Abstract This chapter addresses opportunities and challenges posed by the teaching and learning of mathematics through digital learning platforms basically developed using Moodle (see https://moodle.org). Specifically, we review and discuss the design and implementation of several different mathematics learning environments. Results indicate the existence of new teaching and learning opportunities—and challenges—when working with secondary or middle school students in hybrid learning environments where teaching and learning of mathematics are mediated by technology.

Keywords Hybrid environments · Transforming practices and results at school Online environments · Reflection processes · Resolution of optimization problems

3.1 Purpose of the Chapter

In this chapter, we report on work in mathematics education mediated by technology across four different projects, two of them focusing on technology-mediated and distance mathematics teacher professional development, and two focusing on hybrid mathematics teaching and learning environments at secondary and college

V. Hoyos (✉) · M. E. Navarro · V. J. Raggi · G. Rodriguez
National Pedagogical University, Carretera al Ajusco #24,
Col. Heroes de Padierna, Tlalpan, 14200 Mexico City, Mexico
e-mail: vhoyosa@upn.mx

M. E. Navarro
e-mail: m.estela.navarro@gmail.com

V. J. Raggi
e-mail: vraggic@gmail.com

G. Rodriguez
e-mail: pupitarodriguez@hotmail.com

levels. Through the design and implementation of these teaching and learning modalities, we searched to establish a relationship between student mathematical attainments, functionalities of the different and available digital tools, and the design of the environment. Across these four different projects, we explored the following research questions.

- How is it possible to coordinate syntactic and conceptual aspects of mathematics teaching using specific digital technology in a hybrid learning environment?
- How could student mathematical knowledge be validated through following virtual or online activities?
- How can teacher or student reflection processes be promoted during technology-mediated and distance resolution of math problems?

It should be noted that our interest in the design and implementation of distance learning mediated by technology is based on both a critical look and exploration of the educational potential of this educative modality, when compared to face-to-face mathematics teaching and learning. We also note that many of the main characteristics of the distance or online education seem to challenge theoretical results or notions previously established in the field of classroom mathematics education. For example, years ago, behind the idea of the introduction of new technologies in the classroom was the belief that students would be able to interact and learn with software almost independently of teachers, and with more interaction with knowledge (Sutherland & Balacheff, 1999). However, research has demonstrated the importance of the role of the teacher, by intervening in the negotiation of the meaning of the mathematical activity with the students, to lead them to specific learning (ibidem). Now, in new and massive technology-mediated distance education (for example, see new MOOC's case in Chap. 10 in this book), it has been challenging to model teacher's intervention and/or interaction's orchestration, and/or to promote the exchange of opinions between students on the topic under study, and even in alternative settings it has been generally expected for teachers or students to learn independently (see Hoyos, 2016).

In that way, and according to Balacheff (2010b, 2012, 2015),

Distance learning is provocative because it is a source of restrictions that have their origin in a series of questions and rethinking of common practices.

On the other hand, a noteworthy and common characteristic between all the works that are reviewed in this chapter (see Heffernan, Heffernan, Bennett, & Militello, 2012; Rodríguez, 2015; Hoyos, 2016, 2017) is the existence of significant advances in the understanding and use of digital technologies supporting effectively teaching and learning of mathematics into the school. Specifically, on generalization of patterns in the case of Rodríguez (2015), and on the use of mathematical functions in the cases of Hoyos (2016, 2017) and Hoyos and Navarro (2017). In these works, there is an emphasis on the use of the Internet and/or digital platforms for the administration, storage and deployment of resources, as well as for the interaction between participants and within the resources stored in the platform in use.

Finally, in this chapter, we will share a variety of emerging foci from our research, including the importance of the reflection processes during the resolution of math problems in an online environment (Hoyos, 2016, 2017). Additional foci include equity in schools, specifically in the case of disadvantaged student learning when they were solving complex tasks of school algebra (Rodríguez, 2015), and on improving the classroom environment, particularly in the case of the teaching and learning of functions (Hoyos & Navarro, 2017).

3.2 Hybrid Environments for the Integration of Technology into Mathematics Classroom

Research by Cuban, Kirkpatrick, and Peck (2001), Ruthven (2007), Zbieck and Hollebrands (2008), and Hoyos (2009, 2012) gives an account of few advances on the incorporation of technology into mathematics classroom, and recent OECD data (2015) show that this situation is generalized worldwide. Hence, this section is dedicated to sharing some contributions in this line of research through the design and implementation of hybrid environments for mathematics teaching and learning. The setting, orchestration and exploration of these environments will be presented, including management of distance digital resources (using the Internet in or out of school), and the deployment of the functionalities of specific digital devices in use which, coordinated with the implementation of classroom mathematical activities, have transformed practices and results of mathematics teaching and learning.

3.2.1 Antecedents in the Utilization of Hybrid Environments for Mathematics Teaching and Learning with Technology

Heffernan et al. (2012) reported an investigation on how secondary teachers used ASSISTments (https://www.assistments.org), a *Digital Teaching Platform* (term coined by Dede & Richards, 2012), which was developed by the Worcester Polytechnic Institute to increase, replicate and promote good teaching practices. It includes detailed diagnosis of student misconceptions, provides student immediate specific feedback, and monitors student practice (Heffernan et al., 2012, p. 89). ASSISTments, in the words of these authors, has the characteristics of being:

> a web-based assessment system that provides tutoring based on student response.… Collects data efficiently and provides student-level diagnostic results, … has many features, including content, Mastery Learning, an advanced student-response system, teacher authoring, and data collection.… This is a tool that can be adapted and used in a variety of manners with different cognitive models and content libraries. (Heffernan et al., 2012, pp. 88–89)

A characteristic of ASSISTments, which is important to emphasize here, is related with teachers' hybrid use of the platform, because they have also used it for students to work on out-of-school tasks. Heffernan et al. (2012) have suggested that the components of ASSISTments are designed to highlight what the teacher is already doing in the classroom, but that the use of the digital device allows him or her to do so more efficiently (idem, p. 92).

These authors consider that instructional technologies such as ASSISTments has major implications,

> for the current and future practices of teachers, as for those who direct and train them. If practice is to change to keep pace with the development of new technologies and the expectation of students, then pre- and in-service teacher development efforts must be altered. Moreover, more cloud-based, interactive instructional technologies must be developed and implemented in our schools. (Heffernan et al., 2012, p. 101)

It should be clear up to here that Heffernan and colleagues were not just talking about more cloud-based tech, but those that serve to purposes, such as those highlighted here, students working on out-of-school tasks and teachers practicing more efficiently. The next section of this chapter will present another new research effort, which will be displayed there and that specifically deals with teacher practicing efficiently and innovatively by means of involving hybrid learning environments into the classroom.

To finalize this section, it is important to note that the work of Heffernan et al. (2012) is a significant antecedent of research on digital platforms of teaching not only for inquiring about student achievement, but also because it was established there that this type of research should include more robust, detailed examinations of school level educator development, implementation, and student engagement (Heffernan et al., 2012, p. 101). Having this frame in mind, last section of this chapter will present an investigation by Hoyos (2016, 2017) that speaks partly to this issue, specifically on the examination of secondary teacher professional development on mathematics, at a distance and by means of digital tools.

3.2.2 Advances in Teacher Practice to Innovate by Involving Hybrid Learning Environments

The reviewed work of Heffernan et al. (2012) made in the previous section referred directly to some opportunities that teaching and learning have by means of hybrid environments, providing detailed diagnosis of student misconceptions and monitoring student practice. In these environments teacher participation is central in the planning and orchestration of the use of digital technologies in the classroom—as it is going to be evidenced next, through the review of the work of Rodriguez (2015); thus, increasing the possibility of enabling achievements and reach of mathematical significant ideas. "Significant ideas in mathematics are not necessarily advanced and powerful mathematical notions but instead are key notions that provide real access to the latter" (Rojano, 2002: 144). According to Rojano, significant ideas are those that promote transitional processes and allow students to access levels of thought that surpass specific, numeric and perceptual thinking (Idem). In the review of Rodriguez's work (2015), the reader will recognize the teacher promotion of two transitional processes for students to pass by with the help of virtual manipulatives' tools (see http://nlvm.usu.edu/en/nav/vlibrary.html), namely (1) identification of figural templates (Rivera, 2010); and (2) reproduction and counting of different parts of the figures in a given pattern. Both processes would allow students to access to the generalization of figural patterns, accomplishing then some of the complex tasks that are usual in school algebra.

As part of her doctoral dissertation, Rodríguez (2015) supported seventh grade disadvantaged[1] students in several public middle schools in solving complex tasks on the generalization of figural patterns,[2] a typical subject matter of school algebra.[3] Her pedagogical approach[4] for teaching using hybrid environments for mathematics learning included the use of specific free, online digital materials or virtual

---

[1]According to Dunham and Hennessy (p. 388), disadvantaged students are who traditionally do less well than the general population: "In effect, technology 'leveled the playing field' so that previously disadvantaged groups, who—because of different cognitive styles, learning disabilities, or special circumstances—had usually achieved less, performed as well or better than the main group on outcome measures when using computers or calculators."

[2]Such *figural patterns*, "whether constructed ambiguously, or in a well-defined manner, consist of stages whose parts could be interpreted as being configured in a certain way" (Rivera 2010, p. 298). According to Rivera, he preferred «to use the term figural pattern to convey what I assume to be the "simultaneously conceptual and figural" (Fischbein, 1993, p. 160) nature of mathematical patterns. The term "geometric patterns" is not appropriate due to a potential confusion with geometric sequences (as instances of exponential functions in indiscrete mathematics). Also, I was not keen in using the term "pictorial patterns" due to the (Peircean) fact that figural patterns are not mere pictures of objects but exhibit characteristics associated with diagrammatic representations. The term "ambiguous" shares Neisser's (1976) general notion of ambiguous pictures as conveying the "possibility of alternative perceptions", (p. 50)».

[3]In accordance with Mason and collaborators (1985), a main idea for initiate students in the learning of algebra is that students identify a pattern in a succession of figures (or numbers) and then communicate and record by writing the common characteristics perceived between them, or the relationships that might be established initially with examples. From there teacher can drive some math questions as for example: will there be any formula that could define this pattern? Also, Mason et al., established that once agreed what defines the pattern, the regularities and relationships between its components must be translated from one natural language into a rule or general formula, which will result from a cognitive evolution of the student, such transition is not a simple cognitive exercise, but it could be supported by drawings, diagrams or words, which lead later to describe the key variables in the problem and move to the achievement of its expression in symbolic form.

[4]In his work on a pedagogical approach of teaching, Simon (1995) founded a constructivist learning of mathematics. He developed a model of decision-making for teachers considering the design of math tasks. Its core consists in "the creative tension between the teacher's goals about student learning and his responsibility to be sensitive and responsive to the mathematical thinking of the students" (see Simon, 1995, p. 114). Simon's work presents a diagram (Simon, 1995, p. 136) of a—constructivist—cycle of teaching.

manipulatives hosted by the University of Utah (see http://nlvm.usu.edu/en/nav/
vlibrary.html), for levelling effects in the development of algebraic reasoning of all
students at middle school.

Rodriguez's (2015) work was framed as a teaching cycle[5] (Simon, 1995) on
generalization of figural patterns, using visual templates (Rivera, 2010, 2011) and
utilized specific online digital devices (http://nlvm.usu.edu/en/nav/vlibrary.html) as
the basic tools for the construction of a pedagogical approach to the teaching of the
generalization of figural patterns. Rodriguez's (2015) contribution consisted in
developing digital, interactive, direct and visual manipulation of the Rivera's figural
templates, in order that disadvantaged students could separate, coloring and
counting the distinctive elements of corresponding figural patterns. All these
activities were in fact correlated to specific actions of abduction and/or induction,
actions previously identified by Rivera[6] (2010, 2011) but this time recreated within
the virtual medium, they allowed that participant students developed algebraic
formulas to express the generalization sought.

The empirical observations in Rodriguez's work (2015) were accomplished with
the participation of 45 disadvantaged students from 10 different 7th grade classes.
During 10 sessions of practical work (each session of one hour), each group
(formed by 9 disadvantaged students) was placed in a classroom equipped with
internet-connected computers. Students worked alone in a hybrid environment
basically composed by the use of the virtual manipulatives hosted by the University
of Utah, the following of detailed, printed instructions (or *pedagogical guides,* as
Rodriguez called them) and teacher orchestration of both resources. The work
sessions were video recorded and transcribed, allowing the researchers to construct
detailed descriptions of the student use of the virtual manipulatives, problem
solving, and generalization of figural patterns.

Four hypotheses were advanced in Rodriguez' (2015) research: (H1) First,
according to Simon (1995), a teacher's reconstruction of a (constructivist) pedagogy of mathematics implies that his attention has been able to focus on the
different possibilities of the students. (H2) Second, a mediation using online digital
tools to access complex mathematical ideas will strengthen arguments exposed by
Zbieck and Hollebrands (2008) regarding the benefits of the use of digital technology with disadvantaged students to level their knowledge at the end of a given
schooling. (H3) Third, teacher participation would be central to orchestrate and

---

[5]See last part in the previous final note, in this section.

[6]According to Rivera (2010, p. 300) "meaningful pattern generalization involves the coordination
of two interdependent actions, as follows: (1) *abductive–inductive action on objects,* which
involves employing different ways of counting and structuring discrete objects or parts in a pattern
in an algebraically useful manner; and (2) *symbolic action,* which involves translating (1) in the
form of an algebraic generalization. The idea behind *abductive–inductive action* is illustrated by a
diagram [it appeared in Rivera's work published in 2010, p. 300, in Fig. 5], an empirically verified
diagram of phases in pattern generalization that I have drawn from a cohort of sixth-grade students
who participated in a constructivist-driven pattern generalization study for two consecutive years
(Rivera & Becker, 2008)."

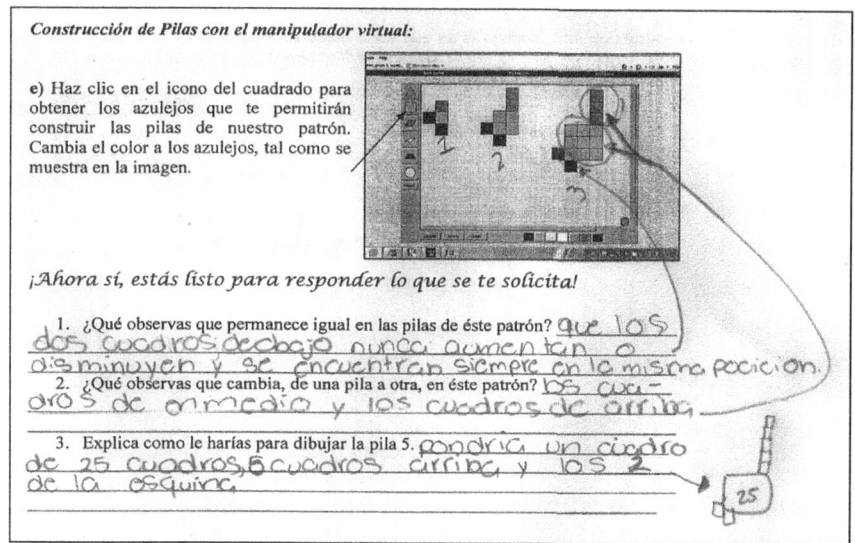

Fig. 3.1 Example of specific instructions and questions given to the students in a printed format, utilized during the empirical observations of Rodriguez's work (2015)

promote student access to specific significant ideas. (H4) Four, the development of mathematical reasoning, particularly in school algebra and with disadvantaged seventh-grade students, would be achieved through the implementation of hybrid learning environments based on a pedagogical reconstruction of mathematics teaching and the mediation of digital technologies, in situations that guarantee the access to significant mathematical ideas and during the resolution of complex mathematical tasks (Fig. 3.1).

The virtual manipulatives used in Rodriguez's study have many of the same features and affordances than concrete manipulatives, which are "objects that can be touched and moved by students to introduce or reinforce an idea or mathematical concept" (Hartshorn & Boren, 1990, Quoted in Neesam, 2015). According to Matus and Miranda (2010), virtual manipulatives have the following characteristics: (i) they tend to be more than the exact replica of the 'concrete', or 'physical', manipulatives; (ii) in general, they include additional options of a digital environment (copy and color pieces, select and move multiple objects); (iii) most offer simulations of concepts and operations that cannot easily be represented by traditional manipulatives; (iv) they are flexible, independent and dynamic; can be controlled entirely by teacher and students; in addition, be used in different lessons, levels and ages; (v) some offer to record the actions or results to provide feedback to the student; and (vi) they are available without limit, anywhere, 24 h a day via the Internet. Teachers, parents and children can often access them for free (Fig. 3.2).

Fig. 3.2 A 7th grade participant student solving one task of generalization of figural patterns during the empirical observations in Rodriguez's work (2015)

The virtual manipulatives hosted by the University of Utah (http://nlvm.usu.edu/en/nav/vlibrary.html) specifically offer the facility for reproducing the figural patterns that appear in the tasks of generalization. This facility allows to move the pieces of the figures into a box to automatically know how many pieces are in a determined subset of the figure, as well as to change the color of subsets according to distinctive figural elements.

For example, in Fig. 3.3, several computer screens show the process of coloring and counting three different subsets or parts of a given figure. This virtual manipulatives' tool is important for students to choose a convenient unity of measure of the change or invariancy that could be perceived in subsequent figures of a figural pattern. This perception (of a unity of measure in figural patterns) is consistent with the identification of visual templates in figural patterns already reported by Rivera (2010). For example, it can be seen in Rivera' research (2010, p. 309) how Emma, one of his students, encircled three subsets in each figure of a given pattern. This was her strategy for counting how subsequent figures were changing. Similar student actions in Rodriguez' study (2015), but now executed

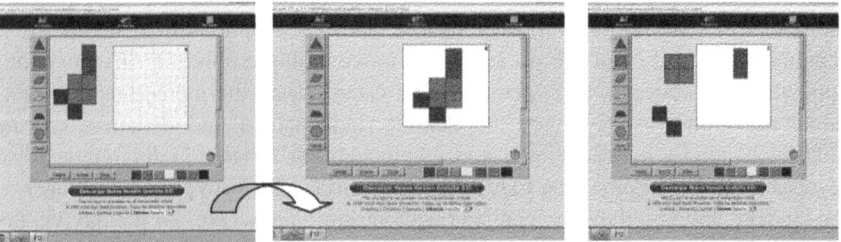

Fig. 3.3 Screens that show the consecutive use of one virtual manipulative, denominated Pattern Blocks, pertaining to the library of virtual manipulatives hosted by the University of Utah (http://nlvm.usu.edu/en/nav/vlibrary.html)

within the virtual manipulatives, are being denoted here by student choosing a convenient unity of measure.

In relation with the tool for counting the number of objects inside the box in the screen, please see Fig. 3.3, the reader could note that the first screen shows an empty box that was made simply by clicking on one side of the figure, and dragging the mouse, holding it. It should also be noted that in the first figure a zero appears in the upper right corner of the empty box. This number is in fact the result of counting the number of elements that have been moved into the box. In the second screen, the whole stack of tiles has been dragged into the box, so the number eight appears on the counter. Finally, in the third screen it is shown that only two elements of the figure have been left inside the box and that the other elements of the figure have been left out. This manipulatives' tool for counting automatically the number of elements in a subset of a figure was beneficial for disadvantaged students, so was the one of coloring the different parts of the figures differently, making both easier for students to identify different subsets of elements (or possible unity of measure in the figures) in such a way that they become important parts of a strategy for registering the changes of consecutive figures in a given pattern.

Perhaps the most important result in Rodríguez' work (2015) is the student's appropriation, using the virtual manipulatives, of the strategy of choosing a specific unity of measure for each pattern, which was partly suggested through the pedagogical approach implemented by Rodriguez, certainly following Rivera's (2010, 2011) work on visual templates. It is to note that once participant students in Rodriguez' work had identified a corresponding unity of measure during a problem, this strategy was subsequently and systematically used, throughout the resolution of subsequent tasks (Fig. 3.4).

Undoubtedly, the pedagogical reconstruction of the topic made by Rodriguez, through building hypothetical learning trajectories (Simon, 1995) for disadvantaged students was a factor related with the success of the activities of resolution in the tasks of generalization of figural patterns with this type of students.

In synthesis, with the help of the virtual manipulatives, disadvantaged students in Rodriguez' study could establish hypothesis about the whole number of elements in any figure of a given (figural) pattern, including those figures that really did not

Fig. 3.4 Two examples of disadvantaged students' executions using the virtual manipulative named as Pattern *Blocks* (see http://nlvm.usu.edu/en/nav/vlibrary.html), and finally at right, an example of a colored figure showing three different subsets and possibilities of choosing a unity of measure

appeared in the diagram included in the text of the problem (abductive action in the process of generalization, in accord with Rivera, 2010); also, they could advance in testing its assumptions (inductive action, in agreement with Rivera, 2010). They could confirm then the validity of the rule they had advanced to define any figure in the given pattern (Rivera, 2010). According to Rivera, both actions (abductive and inductive) are part of the process of generalization, subjacent in the resolution of tasks associated with figural patterns.

In summary, this section has presented the opportunities offered by certain hybrid learning environments for the incorporation of digital technologies into the classroom, as they enable the support by the teacher of students' use of the digital devices. It has been evidenced through the research (Heffernan et al., 2012; Rodriguez, 2015) reviewed in this section, that design and implementation of hybrid environments for mathematics learning has made possible to promote leveling disadvantaged students, and effective and meaningful use by teachers and students of technological tools into the classroom.

3.3 The Challenge of Accomplishing Reflection Processes During Problem Solving in an Online Learning Environment

Two of the questions raised at the beginning of this chapter are related to the content of this third section: How could the mathematical knowledge that students acquired through following virtual or online activities be validated? And, what evidence exists of student reflection during math learning opportunities mediated by technology and/or at a distance?

In this section, we'll explore both questions against the background of an online program (named MAyTE) for the professional development of teachers in mathematics and technology (see Hoyos, 2016). In this work, it was shown that teachers were only able to solve in a procedural way the math problems proposed there (see Hoyos, 2017). One possible explanation for the bias of such type of resolution was that teachers stayed in a pragmatic position when approaching the problems, but they should turn to be able to move to a theoretical one. The hypothesis that is advanced here is that this change of teacher's stance when he is faced to math problem solving will only be achieved through reflection, experimentation of possible solutions and verification or validation of such solutions from an epistemological point of view.

Ninety teachers participated in the six-month online program or course named *MAyTE*, for their professional development (see Hoyos, 2012, 2017), mainly in relation with the incorporation of mathematics technology into their practice, but course focus also included learning to use technology and learning to do mathematics with technology.[7] In this context, it was important to know teachers' strategies during the resolution of complex mathematical tasks and their usage of technology for the resolution, to identify the mathematical resources they displayed, as well as their understanding of the content that were at stake.

In the case of optimization content (a topic from calculus) in college, problems in general are designed for the modelling of real situations. However, the mathematical representations that come into play (e.g. formulas, graphs or symbols, and the treatments or operations carried out with them) obey a set of rules and operative principles within a context of mathematical theories previously established. Thus, when a statement is made in mathematical terms, the validity or not of such statement comes into play—from an epistemological point of view, and this within a well-defined theoretical context (Habermas 1999, quoted in Balacheff, 2010a). Balacheff expresses this complexity of mathematical work as follows: "mathematical ideas are about mathematical ideas; they exist in a closed 'world' difficult to accept but difficult to escape" (Idem).

In the *MAyTE* program, mathematical activities for in-service secondary teachers were developed around an understanding of concepts, learning procedures or mathematical techniques that relied mainly on asking participant teachers in the program for the resolution of some specific mathematical problems, which were designed to challenge secondary teacher math knowledge and teacher implementation of digital tools for resolution, while teachers were only provided with a brief list of instructions and an explanatory text on the mathematical content. *MAyTE* course in general did not include tutorial indications related to the mathematical resolution of the tasks requested.

Balacheff and colleagues' theoretical notion of epistemological validity (e.g. Balacheff 1994, 2004; Balacheff & Sutherland, 1994), and Duval's work on the coordination of representation registers of mathematics, specifically of graphs (Duval, 1994), were useful for analyzing the means or the strategies that participant teachers displayed, using a dynamic software of geometry (*GEOGEBRA* in this case), to solve the problems or learning situations provided in the *MAyTE* program. Years ago, these authors (Balacheff, 1994, 2010a; Balacheff & Sutherland, 1994) illustrated the different contributions certain software has in different virtual learning environments; and here it is noteworthy not only that the teacher

---

[7]Both modes are at the beginning of the incorporation of innovation at the school, according to the PURIA model. Following this model implies that teachers should experiment with the mentioned modes to advance toward successfully incorporating technology into classrooms (Hoyos, 2009, 2012; Zbiek & Hollebrands, 2008).

Briefly, the PURIA model consists of five stages named the Play, Use, Recommend, Incorporate, and Assess modes: "When [teachers are] first introduced to a CAS… they play around with it and try out its facilities… Then they realize they can use it meaningfully for their own work… In time, they find themselves recommending it to their students, albeit essentially as a checking tool and in a piecemeal fashion at this stage. Only when they have observed students using the software to good effect they feel confident in incorporating it more directly in their lessons… Finally, they feel they should assess their students' use of the CAS, at which point it becomes firmly established in the teaching and learning process" (Beaudin & Bowers, 1997, p. 7).

(orstudent) should learn to recognize those different register of representations (Duval, 1994) that are put in play by distinct computational devices or digital tools, in order to solve math problems using these digital resources, but also teachers need to learn about the coordination of representation registers that an appropriate use of computational devices involves when the validation of a solution is in question.

In the case of the resolution of optimization problems, it involves the understanding of two mathematical contents at least, functions and the geometrical relationships at stake; and in according with Duval (1994), "there couldn't be understanding of the content represented without coordination of the representation registers, regardless of the representation register used. Because the peculiarity of mathematics in relation to other disciplines is that the objects studied are not accessible independently of the use of [mathematics] language, figures, schemas, symbols…" (Ibidem, p. 12)

Briefly, in the online *MAyTE* program (see Hoyos, 2012), the activities were developed around the utilization of digital tools that were freely available on the Internet to solve math problems, like *GEOGEBRA*, a software of dynamic geometry. The mathematical content was approached synthetically through a capsule of the content, and the digital tools for solving the mathematical problems consisted of a variety of mathematical software, particularly software of dynamic geometry (SDG).

The text of one of the problems posed is as follows: "A refinery can process 12,000 barrels of oil per day and it can produce Premium [high octane] and Magna [unleaded] gasoline. To meet the demand, the refinery must produce at least 2200 barrels of Premium and 1500 of Magna. The distribution center for the Premium is 30 km from the refinery and the Magna distribution center is from 10 km. The transportation capacity of the refinery is 180,000 barrels/km per day (This means that 180,000 barrels are transported 1 km per day). If the benefit is 20 pesos per barrel of Premium and 10 pesos per barrel of Magna, how many barrels of gasoline should be produced daily to maximize the benefit?"

Most teachers' solutions to this problem were based on the identification and formulation of several algebraic expressions that modelled the given situation, the modelling was in accordance with the data provided, and in the solutions, it was also included a graphical representation using GEOGEBRA, having as starting point the algebraic expressions that firstly were elicited. Such procedures were needed to determine the region of feasibility and the coordinates of the points from which it was possible to obtain the maximum or the minimum cost, depending on the initial conditions of each problem. Many teachers' solutions followed a pattern showed in Hoyos (2016, 2017), it was taken from the documents the teachers uploaded to the platform, and provide evidence of a solution strategy composed of these elements: translation from the initial conditions to algebraic expressions, and representation of these data through the software GEOGEBRA. In this process, the teachers obtained a representation of the feasibility region from which the value of maximum benefit should be deducted. In their graph, the feasibility region was shaded, and the problem in all cases was still unsolved after the graphic was made, because a point [with coordinates (x, y)] needed to be found by means of exploration and through calculating the values of the function of two variables f(x, y), and that in the region of feasibility (for attaining or not the benefit maximum in the case of the problem presented here).

Seeing the image, it should be noted that after having adequately defined the region of points that satisfied the initial conditions, teachers ended by not carrying out an exploration of the f(x, y) values in the region of feasibility, question that would bring them to obtain the requested maximum value, in a point with coordinates that indeed didn't correspond to what visually (see Fig. 3.5) teachers had already chosen or discarded. Readers interested in a detailed account of what teachers did at the time in question, please see it in Hoyos (2016).

Moreover, what is perhaps most interesting is to note that for the computer learning environment in question, in this case constructed mainly for exploration and use of *GEOGEBRA* and for the conversion of mathematical representations (Duval, 1994) required to solve the problem, an epistemological change of teacher's stance must be achieved (Balacheff, 1994, 2004, 2010a), it is linked to the use of the software in the situation or problem proposed, and to the mathematical complexity of the task involved, but basically consist in passing from a pragmatic position (i.e. proposing a solution) to a theoretical one (verifying his proposition).

A way to solve the problem noted in Hoyos (2017) is, for example, by associating any point within the feasibility region to the value of the benefit function, such exploration could thus be carried out directly using *GEOGEBRA*, starting by dragging the point over the feasibility region and verifying the increase or decrease of the value of the benefit function as the chosen point were varying. For example, for point E with coordinates (2507.66, 6006.86) the value of the function f(x, y) equals 110,222 approximately. And it can be proven that the value of maximum benefit is f(x, y) = 149,924.05 when the approximate values for x and y are x = 2993.81, and y = 9004.79.

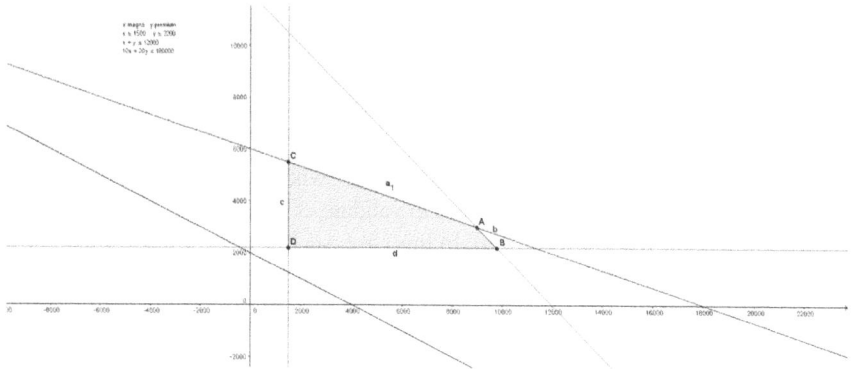

Fig. 3.5 A GEOGEBRA screenshot showing a representation of the feasibility region similar to the image that was part of the teachers' solution in the problem of finding a maximum value of a function f(x, y) in that region

Based on evidence of teacher's stance remained pragmatic during distance problem solving, we concluded that partly of the difficulty to solve could reside into reflect on the possibility of carrying out an exploration using *GEOGEBRA*, starting by dragging a point over the feasibility region and verifying the increase or decrease of the function f as the chosen point was varying. From our point of view, the possibility to do so is entirely relied on a necessity of feedback or teacher (or student) control of their activity within the software (see Balacheff & Sutherland, 1994, p. 15). But this control usually is relied on the coordination of the representation registers or on the comprehension of the mathematical content in question, which is usually not accessed directly by working alone within the software and at a distance.

3.4 Conclusions

The final statement of the last section reminds us, once again, of the great opportunities enjoyed by using hybrid environments for the teaching and learning of mathematics, such as those presented in the first section of this chapter. In these scenarios it has been evidenced, through the review of the work of Heffernan et al. (2012), as an antecedent, and Rodriguez (2015) in this chapter, that teacher's participation is central in the planning and orchestration of the use of the digital technologies into the classroom. For example, in section two of this chapter it has been showed how one teacher was involved in enabling achievements of all students, including those at disadvantage, in the study of complex school subjects, such as in the generalization of figural patterns in school algebra. In fact, this is new evidence on how disadvantage students have benefits rather than deficits when technology is used in supportive teaching and learning environments, as was already noticed by Dunham and Hennessy (2008, p. 375).

By another hand, in relation to the content of the last section (the teaching and learning of mathematics at a distance), principal results drawn from an epistemic and semiotic perspective of analysis (Hoyos, 2017), are as follows: (1) The learning environment was in part defined using a computational device (in this case *GEOGEBRA*) as a procedural tool for the conversion,[8] use and treatment of the different mathematical representations (Duval, 1994, 2006), in this case the equations and graphs that came into play in the given situation of optimization. (2) However, to transit from a pragmatic stance where a possible solution was proposed to a theoretical one that implies to validate or verify it, an epistemological change is required (Balacheff, 2010a). In this case, such change was denoted by the instrumentation of reflective tools, which are not automatically available within *GEOGEBRA* by itself.

---

[8]According to Duval (2006, p. 112): "Conversions are transformations of representations that consist of changing a register without changing the objects being denoted: for example, passing from the algebraic notation for an equation to its graphic representation, ...".

Therefore, based in the evidence and interpretation of data presented in this work and trying to explain teacher difficulties to go ahead towards reflection, experimentation of possible solutions and verification or validation of such solutions in an online PD program and during optimization problem solving, it has been advanced a hypothesis of teacher's necessity of digital collaboration according to his specific participation or activity, or a support to accomplish the epistemological change already mentioned. It would be included in the computational device, or otherwise it would be provided by tutorial intervention (e.g. Balacheff & Soury-Lavergne, 1996; Soury-Lavergne, 1997).

These final remarks mean it is not enough to have access to mathematics technology and/or Internet free resources to achieve expertise or comprehension of certain mathematical content addressed. For example, in the case of teacher resolution of problems of optimization at a distance and mediated by technology, it has been advanced here a hypothesis of necessity of digital collaboration included in the digital device or provided by tutorial intervention to go ahead towards reflection, experimentation of possible solutions and verification or validation of such solutions.

Yet perhaps what is most interesting is that interpreting data from teacher math resolution, framed by Balacheff's & Duval's epistemic and semiotic perspectives, sheds light on how to move forward by correcting the design, incorporating elements missing in the online program reviewed, or working with digital materials as collaborative tools that could promote exploration and reflective thinking to be applied in the solution of certain mathematical tasks, as those that were showed in the situation under study. Finally, this chapter ends with a contribution to the application of the epistemic and semiotic perspectives to reflect upon the potential of using Internet tools and resources for design and research of both distance education of mathematics and math hybrid environments of learning mediated by technology. This reflection can be expressed as follows: In the same way that social interactions do not in principle have an impact on learning but rather depend on the content and forms of interaction chosen, the use of Internet digital tools and computational devices will have an impact on teachers and teaching (or students and learning) when instrumentalization of Internet resources had been exercised to gain knowledge, or to teachers (or students) get control of the activity within the software.

References

Balacheff, N. (1994). La transposition informatique, un nouveau problème pour la didactique des mathématiques, In M. Artigue, et al. (Eds.), *Vingt ans de didactique des mathématiques en France* (pp. 364–370). Grenoble: La pensée sauvage éditions.

Balacheff, N. (2004). Knowledge, the keystone of TEL design. In *Proceedings of the 4th Hellenic Conference with International Participation, Information and Communication Technologies in Education*. Athens, Greece. Retrieved on August 2015 at https://halshs.archives-ouvertes.fr/hal-00190082/document.

Balacheff, N. (2010a). Bridging knowing and proving in mathematics. An essay from a didactical perspective. In G. Hanna, H. N. Jahnke, & H. Pulte (Eds.), *Explanation and proof in mathematics* (pp. 115–135). Heidelberg: Springer. Retrieved on November 2016 at https://arxiv.org/pdf/1403.6926.pdf.

Balacheff, N. (2010b). Note pour une contribution au débat sur la FOAD. *Distances et Mediations des Savoirs (DMS), 8*(2), 167–169). Retrieved on April 24, 2016 at http://ds.revuesonline.com/article.jsp?articleId=15229

Balacheff, N. (2012). *A propos d'un debat provocatrice....* Retrieved from http://nicolas-balacheff.blogspot.fr/2012/06/note-pour-un-debat-provocateur-sur-la.html.

Balacheff, N. (2015). *Les technologies de la distance a l'ere des MOOC. Opportunités et Limites.* Seminar on January 9, 2015, at the National Pedagogical University (UPN), Mexico (Unpublished ppt presentation).

Balacheff, N., & Soury-Lavergne, S. (1996). Explication et préceptorat, a propos d'une étude de cas dans Télé-Cabri. *Explication et EIAO, Actes de la Journée du 26 janvier 1996 (PRC-IA)* (pp. 37–50). Paris, France: Université Blaise Pascal. Rapport LAFORIA (96/33) <hal-00190416> . Retrieved at https://hal.archives-ouvertes.fr/hal-00190416/document.

Balacheff, N., & Sutherland, R. (1994). Epistemological domain of validity of microworlds: the case of Logo and Cabri-géomètre. In R. Lewis & P. Mendelsohn (Eds.), *Proceedings of the IFIP TC3/WG3. 3rd Working Conference on Lessons from Learning* (pp. 137–150). Amsterdam: North-Holland Publishing Co.

Beaudin, M., & Bowers, D. (1997). Logistic for facilitating CAS instruction. In J. Berry (Ed.), *The state of computer algebra in mathematics education.* UK: Chartwell-Bratt.

Cuban, L., Kirkpatrick, H., & Peck, C. (2001). High access and low use of technologies in high school classrooms: Explaining an apparent paradox. *American Educational Research Journal, 38*(4), 813–834.

Dede, C., & Richards, J. (2012). *Digital teaching platforms: Customizing classroom learning for each student.* New York & London: Teachers College Press.

Dunham, P., & Hennessy, S. (2008). Equity and use of educational technology in mathematics. In K. Heid & G. Blume (Eds.), *Research on Technology and the Teaching and Learning of Mathematics: Vol. 1. Research syntheses* (pp. 345–347). USA: Information Age Publishing.

Duval, R. (1994). Les représentations graphiques: Fonctionnement et conditions de leur apprentissage. In A. Antibi (Ed.), *Proceedings of the 46th CIEAEM Meeting* (Vol. 1, pp. 3–14). Toulouse, France: Université Paul Sabatier.

Duval, R. (2006). A cognitive analysis of problems of comprehension in a learning of mathematics. *Educational Studies in Mathematics, 61,* 103–131.

Fischbein, E. (1993). The theory of figural concepts. *Educational Studies in Mathematics, 24*(2):139–162.

Heffernan, N., Heffernan, C., Bennett, M., & Militello, M. (2012). Effective and meaningful use of educational technology: Three cases from the classroom. In C. Dede & J. Richards (Eds.), *Digital teaching platforms: Customizing classroom learning for each student.* New York & London: Teachers College Press.

Hoyos, V. (2009). Recursos Tecnológicos en la Escuela y la Enseñanza de las Matemáticas. En M. Garay (coord.), *Tecnologías de Información y Comunicación. Horizontes Inter-disciplinarios y Temas de Investigación.* México: Universidad Pedagógica Nacional.

Hoyos, V. (2012). Online education for in-service secondary teachers and the incorporation of mathematics technology in the classroom. *ZDM—The International Journal on Mathematics Education, 44*(7). New York: Springer.

Hoyos, V. (2016). Distance technologies and the teaching and learning of mathematics in the era of MOOC. In M. Niess, S. Driskell, & K. Hollebrands (Eds.), *Handbook of research on transforming mathematics teacher education in the digital age.* USA: IGI Global.

Hoyos, V. (2017). Epistemic and semiotic perspectives in the analysis of the use of digital tools for solving optimization problems in the case of distance teacher education. In *Proceedings of CERME 10.* Dublin: Retrieved on 7 August 2017 at https://keynote.conference-services.net/resources/444/5118/pdf/CERME10_0572.pdf.

Hoyos, V., & Navarro, E. (2017). *Ambientes Híbridos de Aprendizaje: Potencial para la Colaboración y el Aprendizaje Distribuido.* Mexico: UPN [in Press].

Matus, C., & Miranda, H. (2010). Lo que la investigación sabe acerca del uso de manipulativos virtuales en el aprendizaje de la matemática. *Cuadernos de Investigación y Formación en Educación Matemática, 5(6)*, 143–151.

Neesam, C. (2015). *Are concrete manipulatives effective in improving the mathematics skills of children with mathematics difficulties?* Retrieved on August 06, 2017 at: http://www.ucl.ac.uk/educational-psychology/resources/CS1Neesam15-18.pdf.

Neisser, U. (1976). *Cognitive psichology.* New York: Meredith.

OECD. (2015). *Students, computers and learning. Making the connection.* Pisa: OECD Publishing. Retrieved on August 7, 2017 at http://dx.doi.org/10.1787/9789264239555-en.

Rivera, F. (2010). Visual templates in pattern generalization activity. *Educational Studies in Mathematics, 73*, 297–328.

Rivera, F. (2011). *Toward a visually-oriented school mathematics curriculum.* Netherlands: Springer.

Rivera, F., & Becker, J. (2008). From patterns to algebra. *ZDM, 40*(1), 65–82.

Rodriguez, G. (2015). *Estudiantes en desventaja resolviendo tareas de generalización de patrones con la mediación de plantillas visuales y manipulativos virtuales* (Ph.D. thesis in mathematics education). Mexico: UPN [National Pedagogical University].

Rojano, T. (2002). Mathematics learning in the junior secondary school: Students' access to significant mathematical ideas. In L. D. English (Ed.), *Handbook of international research in mathematics education.* New Jersey, USA: LEA.

Ruthven, K. (2007). Teachers, technologies and the structures of schooling. En *Proceedings of CERME 5.* Larnaca: University of Cyprus (Retrieved from http://ermeweb.free.fr/CERME5b on 02/15/09).

Simon, M. A. (1995). Reconstructing mathematics pedagogy from a constructivist perspective. *Journal for Research in Mathematics Education, 26*(2), 114–145.

Soury-Lavergne, S. (1997). *Etayage et explication dans le préceptorat distant, le cas de Télé-Cabri* (Thèse du Doctorat). Grenoble: Université Joseph Fourier. Retrieved at https://tel.archives-ouvertes.fr/tel-00004906/document.

Sutherland, R., & Balacheff, N. (1999). Didactical complexity of computational environments for the learning of mathematics. *International Journal of Computers for Mathematical Learning, 4*, 1–26.

Zbiek, R., & Hollebrands, K. (2008). A research-informed view of the process of incorporating mathematics technology into classroom practice by in-service and prospective teachers. In K. Heid & G. Blume (Eds.), *Research on Technology and the Teaching and Learning of Mathematics: Vol. 1. Research syntheses* (pp. 287–344). USA: Information Age Publishing.

Part II
Online Environments and Tutoring Systems for Leveling College Students' Mathematics Learning

Chapter 4
Computer Assisted Math Instruction: A Case Study for MyMathLab Learning System

Adam Chekour

Abstract Colleges and universities are trying alternative instructional approaches to improve the teaching of developmental mathematics with the goal of increasing the number of students who have the skills and knowledge required for college-level math courses and for the twenty-first century workforce. Computers and the Internet make possible new methods of delivering instruction so students will have choices of when, where, and how they learn math. The purpose of this study was to compare academic performance of students enrolled in a developmental mathematics course using traditional instruction and traditional instruction supplemented with computer-assisted instruction. In addition, gender differences in mathematical performance were also investigated. Independent groups T-test was used to compare the mean difference between pretest and posttest mathematics scores of students enrolled in conventional instruction and MyMathLab integrated instruction. Students enrolled in MyMathLab sections made significant gains over students enrolled in conventional sections. Same test confirmed that there was also a significant difference in the posttest scores of females and males, with females outperforming males in both modes of instruction.

Keywords Computer assisted instruction · Hybrid instruction · Conventional instruction · Developmental mathematics · Computer algebra systems MyMathLab · Computer learning systems

4.1 Introduction

Research on mathematics problem solving has largely evolved throughout history from experience-based techniques for problem solving, learning and discovery (Pólya, 1957) to linking these techniques to the development of mathematical

A. Chekour (✉)
University of Cincinnati—Blue Ash College,
9555 Plainfield Road, Blue Ash, OH 45236, USA
e-mail: adam.chekour@uc.edu

© Springer International Publishing AG, part of Springer Nature 2018
J. Silverman and V. Hoyos (eds.), *Distance Learning, E-Learning and Blended Learning in Mathematics Education*, ICME-13 Monographs,
https://doi.org/10.1007/978-3-319-90790-1_4

content (Lester & Kehle, 2003). Exposing students to the course content has often not been enough for them to achieve academic success in mathematics. Implementing a variety of instructional strategies that increase students' motivation and meaningful learning were also necessary. Only recently, math problem solving has known an infusion of a variety of technology tools and procedures aimed at enhancing students' meaningful understanding of different math concepts (Lesh, Zawojewski, & Carmona, 2003).

Indeed, research on how mathematics is integrated in different fields (e.g. STEM education), and how professionals in these fields tend to heavily apply mathematical concepts, has dramatically affected the nature of math problem solving. Furthermore, this integration emphasizes the necessity of implementing new and powerful technologies to enable students' conceptualization, communication, and computation while solving math problems (National Council of Teacher of Mathematics, 2000). These skills certainly provide students with a new perspective on how to approach math problem solving and build a foundation for them to be a successful critical thinker and problem solver within and beyond school (NCTM, 2000).

However, the implementation of technology in the teaching and learning of mathematics has witnessed a slow growth, due to factors such as accessibility to technological tools, students' and teachers' beliefs about technology, and lack of general research enlightening the effects of technology on enhancing classroom instruction, mathematical curriculum content, and students' learning of mathematics (Zbiek, Heid, Blume, & Dick, 2007). Such an issue persists mainly in K–12 education and developmental mathematics classes at two-year or community colleges (International Technology Education Association, 2006). While there are numerous case studies on specific technologies applied to K–12 math education, there is still a need for a comprehensive synthesis of the findings of these separate case studies. This will inform, substantially, both the practice and the research in math education.

Further, most of the research studies on technology-infused math education emphasizes only the technical aspect of learning mathematics, which involves mathematical activities and procedures that lead to numerical computations, solving equations, using diagrams, and collecting and sorting data (Borwein & Bailey, 2005). Conversely, few research studies address the instrumental use of technology to enhance students' conceptualization of math activities involving how students understand, communicate, and use mathematical connections, structures, and relationships. Achieving this goal not only changes educators' and students' belief about technology, but it also improves students' skills in learning mathematics (Kulik & Kulik, 1991).

Although problem solving has been a concern of psychologists and education researchers for more than a hundred years, there have been limited research studies on mathematical problem solving and math reasoning involving the use of technology. In addition, few methods have been implemented to study the various concepts of problem solving, such as simulation, computer modeling and experimentation (Maccini, Mulcahy, & Wilson, 2007). One aspect of computer modeling is the use of Computer Algebra Systems (CAS) such as the MyMathLab application, which has found its way to a variety of math course levels and is currently adopted by numerous academic institutions (Pearson Education, 2005). Initially designed and commercialized by Pearson Education, a worldwide leader in education resourcing, MyMathLab has been, somewhat, a successful tool in enhancing students' learning, specifically in developing math problem solving skills (Pearson Education, 2005).

This research is intended to evaluate the success of implementing MyMathLab into the learning process while solving math problems and learning developmental mathematics. The efficacy of Computer Assisted Instruction (CAI) using MyMathLab will be compared to traditional, face-to-face instruction of mathematics in developmental classes.

4.2 Significance of the Study

High school graduates often come unprepared to college math courses and therefore struggle in meeting math course expectations (National Center for Educational Statistics, 2003). Despite the continuous intervention efforts of different institutions, only 10% of these students graduate, and only 40% of these students who are in developmental math programs graduate with a bachelor's degree (Brittenham et al., 2003).

Whether the developmental math classes will lead students to attend four-year institutions, with more emphasis on college algebra and statistics (NCES, 2003), or to qualification for a meaningful job, colleges and universities are concerned with a low passing rate not even exceeding 24% at some colleges (Trenholm, 2006). Therefore, there is urgency in developing programs and strategies that aim at student retention and provide a meaningful learning experience to students, one that emphasizes understanding of math concepts, promotes active constructivist learning, and allows for transfer to real world applications. Instructors of developmental mathematics are implementing different supplemental tools to traditional instruction known to be limited in effective resources and pedagogies. The broadened use of computer technology in education today has led math instructors to implement computer tools to benefit students' learning of mathematics (NCES, 2003). The National Council of Teachers of Mathematics (NCTM, 2000) calls for using computer technology as a means to enhance math teaching and learning in and out of classrooms. In addition, in a study conducted by the National Center for Educational Statistics (2003), 31% of the 3230 US surveyed colleges (during the fall of 2000) revealed a frequent use of computers by students for their instructional needs in on-campus remedial math education.

With a fast increase of computer technology, a variety of software, hardware and media tools has found its way into developmental mathematics to offer a rich learning experience to students, while they are learning mathematics. Most of the software used in developmental mathematics has been developed by textbook publishers to either supplement classroom instruction with tutorial and algorithmically generated problems, or to provide a thorough presentation of concepts with interactive multimedia (Kinney & Robertson, 2003). In addition to the advantage of receiving immediate feedback, students can also self-pace and revisit their assignments until mastery (Hannafin & Foshay, 2008), although there is a schedule for completion of lessons. They can also benefit from accessing a variety of built-in resources such as videos, guided practice problems, and online tutoring. Teachers can also build individualized study plans, quizzes and tests immediately graded by the software and tailored to a specific unit and learning objectives.

Providing instructors with detailed data on students' progress is a valuable feature to course overall assessment (Cotton, 2001). Within this perspective, this study aims at investigating the effect of computer-assisted instruction on the mathematical learning of students in developmental classes, using the MyMathLab learning system. The results of this study can inform institutions in investing their resources wisely on computer-assisted instruction with potential impact on students' mathematical achievement. The study also suggests future research to decipher key student characteristics that are associated with higher developmental math achievement, within different delivery formats, and simultaneously, improve the experience of computer-assisted learning of mathematics in developmental courses.

4.3   Theoretical Framework

There are two major theories that provide the framework for analyzing the data and guiding the discussion in this study. The first theory is constructivism grounded in eminent work of Dewey (1958) and Vygotsky (1978). The main premise of constructivism is the learner's ability to internally construct an understanding of mathematical concepts and connect them through important relationships. This constructivist learning usually conditionalizes knowledge through experience, exploration, thinking, and reflection (Dewey, 1958). In addition, this experiential learning often takes place within an interactive environment, which promotes understanding through life experiences, or can be mediated by an educator, who usually guides this discovery process.

The second theory is grounded in the nature of technology use, which categorizes technology into instrumental and substantive (Trenholm, 2006). According to Zbiek, Heid, Blume, and Dick (2007), the instrumental view of technology is a legacy of Aristotle, who posited that technological products have no meaning in and of themselves and that technology receives its justification from serving human life. Indeed, technology use in mathematics will have no meaning if it doesn't promote or supplement the learning process of students. At the same time, the use of technology needs to maintain the right balance and fit the right purpose, by allowing a deeper-level of logical and critical thinking.

The substantive view of technology is based on the view that "technology is becoming autonomous, is going beyond human control, and operates and evolves according to its own logic and norms" (Chen et al., 2011, p. 57). This theory recognizes that the learner's experience is mediated by and structured through technology. Indeed, students and technology become embodied in an experientially semi-symbiotic unity, where the technology mediates what and how mathematics is experienced and learned. In this process, the technology becomes appropriated to the learning activity as an integral part of students' thinking, causing an inter-weaving of instrumental and mathematical knowledge, and preventing learners from accessing the types of activities and mathematics that are not afforded by the particular technology.

4.3.1 Issues in Math Teaching and Learning

While students are required to acquire the ability to compute, problems solve, and put mathematical concepts and skills into practice, to compete with the demands of a fast growing and technology saturated world, there are still several increased challenges facing the learning and teaching of mathematics. The United States Department of Education indicates that U.S. students are performing below their counterparts in other developed countries (USDOE, 2008). In addition, the National Assessment of Educational Progress (2006) claims that only two percent of U.S. students manage to attain advanced levels of math achievement by grade 12.

Woodward (2004) contends that changing policies and standards in mathematics are among possible solutions for the decline of student performance in mathematics. Indeed, the National Council of Teachers of Mathematics (NCTM) initiated reform efforts to improve students' achievement in mathematics through revising math curricula and core standards. This revision process focused primarily on a thorough and a deeper pedagogical content knowledge of conceptual understanding (NCTM, 2000). The focus highlighted a significant shift from the rote memorization of computational facts and procedures used in math problem solving to practical situations, where students are given opportunities for critical thinking and problem solving.

Another issue pertaining to students' performance is the lack of effective instructional approaches and metacognitive strategies, which aim at enhancing and scaffolding the mastery of abstract concepts in mathematics (National Mathematics Advisory Panel, 2008). According to Maccini and Gagnon (2002), authentic mathematics instruction should include the following recommended instructional practices:

1. Differentiated instruction
2. Metacognitive strategies and instructional routines
3. Progress monitoring and formative assessment procedures
4. Computer-assisted instruction and Universal Design (p. 13)

In another study, Bouck and Flanagan (2009) claims there are two fundamental constraints hindering the improvement of K–12 mathematics education. The first one lies in the fact that mathematics teachers spend valuable energy and time in designing instructional activities, which are seldom conductive to students' exploration and construction of their own understanding of mathematical concepts. The second constraint is that mathematics instructors are resistant to changes in their teaching strategies. Indeed, Bouck claims that the current teaching of mathematics is mainly characterized by universal formal and symbolic presentations of mathematical rules or procedures, based on a mere textbook presentation rather than a synthesis of encompassing mathematical relations.

In a subsequent study, Kilpatrick, Swafford, and Findell (2001) maintains that the most fundamental task facing mathematics instructors is to promote mathematics conceptual meaning among their learners. This approach supports the Piagetian position that postulates the existence of mathematical objects as synonymous with meaning, which later become mental entities that unfold over time to allow for connection and transfer of knowledge (Piaget & Garcia, 1991).

In a different study on the use of constructivism in math activities, Lerman (2001) argues that conceptual knowledge cannot be transferred automatically from teacher to learner, but it should be built and molded by the learner under the guidance of the instructor, and on the basis of prior knowledge and experience. According to Lerman, this constructivist representation of one's reality and meaning enables learners to seek the meaning in the structure, organization, and relationships between different math subjects.

Within the same perspective, Santos-Trigo (2007) contends that mathematics learners achieve better outcomes in problem solving when they assimilate mathematical concepts to the collection of their intrinsic satisfying models. The absence of these intrinsic models makes mathematical problem solving painfully difficult, even when concepts are simulated. This is due to a lack of insights, which mystically shed the light on and facilitate the process of problem solving.

In her study on scaffolding problem solving, Muis (2004) claims that the socialization of problem solving through discussion is essential to the success of this process. This socialization suggests a shift in the focus of activities led by the instructor to activities lead by the students, but understood and facilitated by the former (Kim, 2011). This provides the instructor with a context to be more sensitive to students' mathematical experience, in addition to developing meaningful mathematical conversations conductive to concepts' assimilation, organization, extension and transfer.

Within the same construct, Magiera and Zawojewski (2011) contend that the lack of meaningful mathematical discussions is frequently observed in mathematics classrooms. Students tend to work independently on math tasks and activities, without benefiting from the opportunities to communicate and interact with their peers. The socialization aspect of these mathematical conversations is fundamental to critical thinking, problem solving and evolving as a mathematician. Therefore, math instructors need to perpetuate these behaviors and make them a common place in the mathematics classrooms.

Recently, a continued emphasis was made on mathematics and science integration to improve teachers' knowledge in both mathematics and science. According to Stinson, Harkness, Meyer, and Stallworth (2009), math integrated instruction enables students to better understand problem-solving strategies as they are applied to more practical situations. Instructors also benefit from math integration by experiencing greater student involvement and contribution to the design of effective math curricula (Stinson et al., 2009).

Finally, within the same perspective, Schoenfeld (2007) maintains that mathematics instructors should possess two kinds of competencies, subject matter knowledge and general pedagogical skills, in order to achieve satisfactory and efficient teaching. The intertwining of these two competencies constitutes teachers' belief and affects towards mathematics and mathematical activities, which also affects students' potentials for learning mathematics (Schoenfeld, 2007). Therefore, it is important to identify and implement professional development components that are specific and instrumental for pre-service preparation, in-service development, and professional identity in the field of mathematics education (Sfard, 2006).

4.4   Methodology and Sampling

4.4.1   Methodology

This is a quasi-experimental study that uses a non-randomized pretest-posttest design. Students will self-register to class, with no disruption of their schedules. The target population consists of several developmental math classes selected from a Midwestern university, which will provide the sampling convenience. One of each paired classes received traditional instruction (control group), and the other received traditional instruction supplemented with the MyMathLab computer learning system (treatment group). Both groups were subjected to a pre-test and a post-test. The pre-test was a math placement test students generally take to be placed in a certain math course level, while the post-test (also consisting of math placement test) was given as a review test to the course's final exam. Instruction of both groups was coordinated with the content of the textbook. In this study, samples might not represent the real college population, but generalizability was attempted through sampling different courses and their sections, yet presenting similar characteristics of developmental classes.

The independent variable of this study is the mode of instruction, with the entrance math placement test as the pre-test. The dependent variable is the student's math performance, as measured by the math department placement test scores at both pretest and posttest, in all courses. Before the beginning of fall term 2014, both control and experimental groups took a pretest as an academic requirement to register to any developmental mathematics course. A week after the start of the course, all students completed an online questionnaire, the purpose of which was to gather descriptive data to establish similarities of both groups. Students were placed in developmental mathematics class based on their math placement test scores (pretest). Therefore, all students were expected to have mathematics achievement within the same range of scores. Since students self-registered for the classes, they couldn't be placed by the college into a section with a particular mode of instruction. However, students were having access to information about which sections will use computer-assisted instruction and which will use traditional instruction from their developmental mathematics instructors, their advisors, the developmental math website, and the tutoring center. The instruction was delivered for 16 weeks, and during the last week, all students took a version of the placement test prepared and reviewed by the math department faculty. Data will be compiled regarding instructional method, and pretest and posttest scores. The questionnaire results will be gathered to serve as information on students' demographics.

An analysis of variance was conducted using both T-test and ANOVA. The independent variable (IV) was the method of instruction, and the dependent variable (DV) was the math performance measured by the math placement test (at both the pretest and posttest). This analysis was intended to check the first null hypothesis that states: There is no significant difference in the mathematics performance of students in a developmental mathematics course using traditional instruction and computer assisted instruction (traditional instruction supplemented by MyMathLab learning system).

The second null hypothesis stated that there is no significant difference in the mathematical performance of developmental mathematics students by gender. A second analysis of variance using both T-test and ANOVA was conducted with gender as the independent variable and the posttest as the dependent variable. The interaction of method and gender was also analyzed. Both groups were taught by full-time instructors who have demonstrated competence in teaching developmental mathematics students and in teaching in both delivery formats (face-to-face and hybrid).

4.4.2 Study Participants

The responses to a demographics questionnaire were expected to reveal some similarities, consistent with national studies, which report that 56% of undergraduate students in 2000 were female (NCES, 2005), and 55% of community college remedial students were female (Boylan, 2002).

The responses to the questionnaire revealed some noticeable similarities between students in the traditional and traditional with computer assisted instruction classes. Students of both groups were more likely to have the same age, to be white, to have few years since their last math course, and to have positive attitudes toward math and computers. Three-hundred-seventy-one students had taken their last math class one or two years ago.

Overall most students had been in college two to four quarters. All of the students were full-time students, taking at least 12 credit hours.

Thirty-five percent of the traditional students, and 28% of the traditional + CAI students, reported having a negative or very negative attitude toward math. Nearly all students (352 of the 371) used computers for both academic purposes and other reasons, such as email, social networking, and shopping. Seventy-one percent of the traditional students, and 59% of the traditional + CAI students reported feeling positive or very positive toward using computers for educational purposes.

Overall, 44% of the students were male and 56% were female, which is consistent with national studies that reported 56% of undergraduate students in 2000 were female (NCES, 2005), and 55% of community college remedial students were female (Boylan, 2002). While, 95% of the students were of traditional college age (less than or equal to 23). Overall, 74% of the students were white, 14% were African American, 4% were Hispanic, and 8% were other ethnicities.

4.4.3   Instrumentation

The construct of mathematics achievement was operationally defined as scores on the developmental mathematics placement test, at both the pretest and the posttest. This is the departmental placement test given to all students in developmental mathematics. Validity of this test is the extent to which it measures mathematics achievement. The test is a collection of test exam items created by the developmental mathematics faculty and matched with the developmental mathematics course objectives in proportion to the emphasis given to each topic during the semester. This provides face validity. The test questions are reviewed and critiqued by a team of Department of Mathematics faculty members. This provides content validity. Reliability of the placement test is the extent to which scores are free of random error, that is, the extent to which the exam yields consistent results. Ideally, the reliability coefficient should be close to one. Cronbach's Alpha, or coefficient alpha, for the final exam is 0.915 as calculated using SPSS, based on scores from a sample of 100 exams from eight instructors from previous semesters. Cronbach's alpha calculates the mean of all possible split-half correlations and is preferred by many researchers when the questions have different point values, such as a Likert scale or essay test (Ary, Jacobs, Razavieh, and Sorensen, 2006).

The pretest consists of 30 questions from the placement test, representing major course objectives of developmental mathematics. The placement test as the pretest was taken by all students as an academic requirement to test into the course. The test usually requires 2 h to be completed. The placement test as the posttest consisted of 30 questions again and was given few days before the final exam, as part of the course requirements, within a 120-min exam period. The pretest serves to inform the students that they did indeed need to take the course and gives them a preview of topics that are usually studied in developmental mathematics. Cronbach's alpha for the pretest was also 0.915.

The nonrandomized control group, pretest-posttest design does not provide the control that a randomized experimental design does because subjects are not randomly assigned to groups (Ary et al., 2006). The more similar the control and experimental groups are at the beginning of the experiment, as determined by the questionnaire and similar means on the pretest, the more credible are the results of the study. Threats to internal validity were controlled where possible. An analysis of variance was conducted on the pretest and posttest scores, and the questionnaire data established the similarity of the groups before treatment.

Attitudes of the subjects toward mathematics or technology may affect the outcome of an experiment (Ary et al., 2006). In this study the effect of attitudes was controlled by not telling the subjects they will be participating in a study. Many developmental mathematics students, whether they participated in the study or not, completed a questionnaire (see Appendix), and all have taken the mandatory pretest and posttest. Extraneous variables were controlled where possible.

All groups had the same course objectives, same schedule, same tests, and the same 16 weeks of instruction. Each instructor was teaching in both modes of instruction. Attrition might have been a threat as more students with low scores withdrew from one group than the others.

4.5 Data Analysis and Narrative of Findings

Table 4.1 shows the distribution of the 371 participants as per method of instruction and gender. A total of 371 students took the pretest: 187 from the traditional (control) and 184 from the integrated traditional/MML (intervention); the same number of students took the posttest. Of this group, 146 were male and 255 were female. Therefore, 79 male and 108 female students were enrolled in the control group, and 83 male versus 101 female students were enrolled in the intervention group (Table 4.1). Of the total 172 participants, 162 (44%) were male and 209 (56%) were female, with an almost even split between traditional instruction and traditional instruction supplemented with MyMathLab learning system.

As both t-test and ANOVA were implemented in this study, with two independent variables (mode of instruction and gender) at two levels each (traditional versus hybrid instruction and male versus female), several assumptions must be met. The first assumption involves the instrumentation scale of measurement: data collected for the dependent variable should be continuous or ordinal. The instrument used for reporting test scores in this study meets this assumption.

Table 4.1 Distribution of sample by method of instruction and gender

Method	Male	Female	Total
Traditional	79	108	187
Traditional + MML	83	101	184
Total	162	209	371

The second assumption, random sampling, was impossible to satisfy given that students self-registered to classes; however, paired sections were compared to determine similarity of demographics and pretest scores. The third assumption pertains to testing for normality using skewness and kurtosis values (D'Agostino, Belanger, & D'Agostino, 1990). This test was conducted on the dependent variable posttest (MPTPOST), for both subgroups (traditional versus hybrid and male versus female). Both Kolmogorov–Smirnov and Shapiro–Wilk tests (Tables 4.2 and 4.3) infer a normal distribution for the dependent variable MPTPOST in each method of instruction ($p < 0.05$). This is confirmed by the Q-Q Plots (Figs. 4.1, 4.2, 4.3 and 4.4), where the value for each score is plotted against the expected value from the normal distribution. The reasonably straight line for both male and female plots suggests a normal distribution for the MPTPOST dependent variable. Similar results are shown when examining gender, thus indicating a normal distribution. The fourth assumption is that of adequate sample size. Each group (traditional versus hybrid, and male versus female) has more than 30 cases and is therefore sufficient for the analysis.

4.5.1   Hypothesis I

Hypothesis I asks whether there is a significant difference in the performance of students in developmental math classes based on type of instruction. To begin the analysis, an independent groups t-test was conducted to determine whether there was a significant difference between the pretest means of the control group versus the intervention group. Results indicated no significant difference between groups ($p = 0.003$) (Table 4.2). The mean pretest score for students registered for

Table 4.2 Independent samples T-test results of the post-test by mode of instruction

Mode of instruction	F	Levene's test for equality of variances		t-test for equality of means			
		Sig	t	df	Sig (2-tail)	Mean diff	Std error diff.
Equal var. assumed	53.9	0.003	−5.53	369	0.001	−11.82	−16.03
Equal var. not assumed			−5.54	314	0.001	−11.82	−16.02

Table 4.3 Descriptive statistics by method of instruction

Method	N	Pretest		Posttest	
		Mean	SD	Mean	SD
Traditional	187	69.87	29.23	96.02	30.13
Traditional + MML	184	72.31	29.94	129.76	28.08
Total	371	71.35	29.67	112.66	29.56

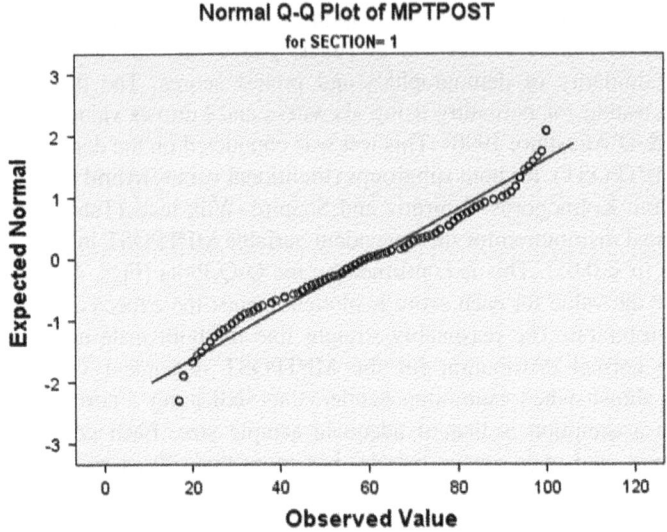

Fig. 4.1 Q-Q Plot of normality of posttest by mode of instruction (control group)

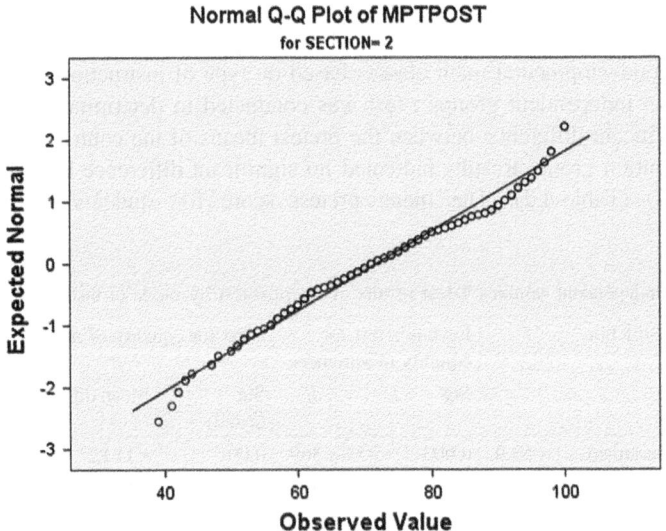

Fig. 4.2 Q-Q Plot of normality of posttest by mode of instruction (intervention group)

traditional instruction was 69.87 (SD 29.23). The mean pretest score for traditional + MML students was 72.31 (SD 29.94). The total mean score was 71. (SD 29.67). Because there was no significant difference in mean pretest scores between the control and intervention groups, analysis of the difference in pretest

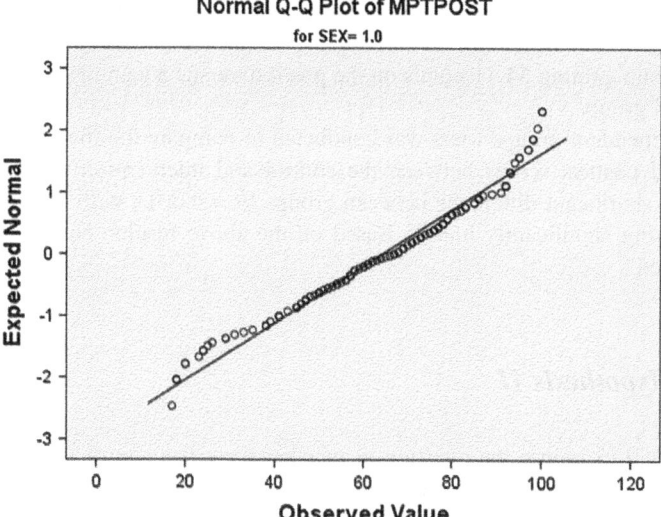

Fig. 4.3 Q-Q Plot of normality of posttest by gender (males)

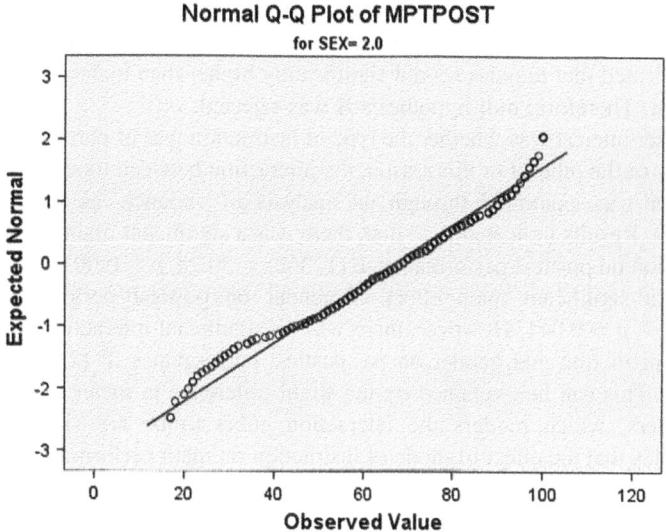

Fig. 4.4 Q-Q Plot of normality of posttest by gender (females)

and posttest scores between both groups could proceed without the need to adjust for pretest values. Results (Table 4.3) also indicate that the mean posttest score for the control group was 96.02 (SD 30.13), while the mean posttest score the intervention group was 126.72 (SD 28.08). The total mean score was 112.66 (SD

29.56). Unlike the difference between group means on the pretest, these results show a significant difference between group means ($p = 0.001$), with the intervention group gaining 54.41 points on the posttest versus a gain of 26.15 points for the control group.

An independent groups t-test was conducted to compare the mean difference in pretest and posttest scores between the control and intervention groups. Results revealed a significant difference between groups ($p = 0.001$), with the intervention group scoring significantly higher. Based on the above results, Null Hypothesis I was rejected.

4.5.2   Hypothesis II

Hypothesis II examines the question of whether there is a significant difference in achievement between males and females in developmental math classes. Of 371 students who completed the pretest and posttest, 162 were males and 209 females. As indicated in Table 4.4, the mean posttest score for males was 65.33 (SD 21.72), while the mean posttest score for females was 68.86 (SD 21.19). For all 371 participants, the mean posttest score was 66.26 (SD 21.41). An independent group t-test was conducted to determine whether there was a significant difference between the mean posttest scores of males and females. According to Table 4.5, results indicated that females scored significantly higher than males ($p = 0.028$) on the posttest. Therefore, null hypothesis II was rejected.

Of further interest was whether the type of instruction was of particular benefit to one gender or the other. For this reason, the interaction between mode of instruction and gender was examined through an analysis of variance, as summarized in (Table 4.6). Results indicated that, first, there was a significant main effect of mode of instruction on posttest performance, $F(1, 368) = 30.78, p = 0.002$. Second, there was also a significant main effect of gender on posttest performance, $F(1, 368) = 7.67, p = 0.041$. However, there was no significant interaction between the mode of instruction and gender on the posttest performance, $F(2, 368) = 52.01$, $p = 0.166$. This can be explained by the slight difference in mean values between both genders, which renders the interaction effect to be non-significant. This demonstrates that the effect of mode of instruction on math performance scores was not different for male participants versus females.

Method	N	Pretest		Posttest	
		Mean	SD	Mean	SD
Male	162	56.71	23.51	65.33	21.72
Female	209	58.25	24.56	68.86	21.19
Total	371	57.64	24.13	66.26	21.41

Table 4.4 Descriptive statistics by gender

Table 4.5 Independent samples t-test results of the post-test by gender

Mode of instruction	F	Levene's test for equality of variances		t-test for equality of means			
		Sig	t	df	Sig (2-tail)	Mean Diff	Std Error Diff.
Equal var. assumed	0.413	0.041	−1.111	369	0.267	−2.5266	2.2748
Equal var. not assumed			−1.12	304.31	0.028	−2.53	2.29

Table 4.6 Interaction between mode of instruction (section) and gender (sex)

Source	Type III SS	df	Mean square	F	Sig
Corrected model	14,365.40	4	7182.70	17.02	0.001
Intercept	207,129.58	1	207,129.58	30.78	0.002
Section	12,987.69	1	12,987.69	30.78	0.002
Section * Sex	346.19	2	346.19	52.01	0.166
Error	155,271.13	368	421.93		
Total	1,779,251.00	371			
Corrected total	169,636.53	370			

4.6 Discussion of Findings

The major finding of this study was the superiority of instruction that integrated MyMathLab over traditional instruction in college developmental math classes. Students in the integrated sections scored 30.7 points higher on the posttest than students in the traditional sections, although both groups had comparable pretests scores. The posttest scores of the sections that integrated MyMathLab indicate that the students are now prepared to enter regular college math courses, while the students who participated in the traditional sections continue to need more developmental math courses. Based on the results of this study, colleges should move to incorporate MyMathLab into their developmental math courses.

A secondary finding that is of interest to instructors of developmental math courses is the performance of females in this study. Regardless of method of instruction, females scored higher than males at posttest. These results indicate the females have bridged the gender gap in terms of performance in mathematics, at least at this level.

4.7 Study Implications

Developmental mathematics students are unprepared for college-level mathematics courses and need a learning experience that is different from their learning experiences in middle and high school—experiences that resulted in them being placed in a developmental course in college. Since developmental students are very diverse in mathematical background and have a variety of learning styles, no one instructional style will meet the needs of all students. Therefore, colleges and universities should offer developmental mathematics courses in a choice of instructional models. The findings of this study indicate that developmental mathematics students learn better from a lecture supplemented with computer-assisted instruction, rather than from a lecture alone.

Several reasons indicate that colleges and universities should offer developmental mathematics courses with computer-assisted instruction. Standards developed by the American Mathematical Association of Two-Year Colleges (1995) and the National Council of Teachers of Mathematics (2000) call for the use of technology in the classroom to improve student learning. Technological advances have made computers more powerful and less expensive, which has resulted in more students having access to computers. Most college students are inclined to use them for academic purposes in addition to communication and social uses. Eighty-five percent of college students in the Pew Internet and American Life Project (2002) had their own computer and 79% said the computer had a positive impact on their college academic experience. Finally, much of the research indicates that students of all ages and abilities using computer-assisted instruction in a variety of instructional models learn as well or better than those receiving traditional instruction.

Developmental educators should learn how to use technology effectively to improve student learning. One of the factors identified as critical to success in an online developmental mathematics course was professional development for faculty (Perez & Foshay, 2002). A report based on 176 literature reviews and individual studies found that the achievement of students using computer-based instruction was significantly related to the amount of technology-related training the teachers had received and whether the technology was being used appropriately (Bialo & Sivin-Kachala, 1996).

Faculty should constantly evaluate computer software because new products continue to be developed and old ones change. Some software is designed to supplement classroom instruction and some is designed to deliver instruction (Kinney & Robertson, 2003). Instructors need time to evaluate and select software appropriate to the course design. They need to know how to use the technology and how to integrate it in the curriculum in a way that enhances student learning. Since developmental students often lack study skills, organizational skills, and motivation (Armington, 2003); courses with an online component should include lessons and discussion boards on learning strategies (Kinney & Robertson, 2003; Trenholm, 2006; Wadsworth, Husman, Duggan, & Pennington, 2007).

In order for students to receive the maximum benefit from using a computer learning system, faculty should provide instruction in how to use the system. Researchers have discovered a high degree of frustration among students and teachers in communicating with mathematical symbols (Engelbrecht & Harding, 2005; Jacobson, 2006; Smith & Ferguson, 2004; Testone, 1999). Students need to learn how to enter mathematical notation. A student may have the correct answer on paper but the computer will not accept it as correct if the answer is entered improperly. They also need to know how to use the tutorial features and the study plan to improve their learning. Some students attempt the graded assignments without first working the tutorials and become discouraged when they earn low scores. Students should also be taught how to monitor their progress in the course using the grade book.

This study also indicates that some developmental mathematics students do learn in a traditional classroom. Although lecture alone has not been effective with developmental students, there is evidence in the literature that enhancing the lecture with such techniques as group work, cooperative learning, class discussions, real-world examples, and peer tutoring has positive results. Educators using the traditional lecture should examine their teaching practice and find ways to enhance the lecture with active learning and relevant examples that will motivate students to learn. Courses could be redesigned with classes meeting four or five days a week. Two or three days could be lecture and the remaining days would be for students to work problems and take quizzes.

Developmental educators should strive to give all students, whether male or female, equal opportunities to receive a quality education. Instructors should examine whether they treat males and females differently in any way, including asking and answering questions from one gender more than the other, and then make necessary corrections. A peer or supervisor could conduct a classroom observation in which the number and types of interactions are recorded by gender.

4.8   Conclusion

Based on the literature and the findings of the current study, several conclusions can be drawn concerning developmental mathematics and computer-assisted instruction. The results of this study indicate that developmental mathematics students learn better with computer-assisted instruction (such as MyMathLab learning system) than with traditional mode of instruction. The mere presence of computers does not improve student learning, unless used carefully. Students have an interest in using technology for a variety of purposes including academics. Computers still have the potential to be useful tools to improve learning. They provide educators the opportunity to create courses in a variety of alternative formats to the traditional lecture in order to address the different learning styles and preferences of students. Quality is essential in any mode of instruction.

References

American Mathematical Association of Two-Year Colleges. (1995). *Crossroads in mathematics: Standards for introductory college mathematics before calculus.* Memphis: Author.

Armington, T. C. (Ed.). (2003). *Best practices in developmental mathematics* (2nd ed.). NADE Mathematics Special Professional Interest Network.

Ary, D., Jacobs, L. C., Razavieh, A., & Sorensen, C. (2006). *Introduction to research in education* (7th ed.). Belmont, CA: Thomson & Wadsworth.

Bialo, E. R., & Sivin-Kachala, J. (1996). The effectiveness of technology in schools: A summary of recent research. *School Library Media Research, 25*(1), 57 (Fall).

Bouck, E. C., & Flanagan, S. (2009). Assistive technology and mathematics: What is there and where can we go in special education. *Journal of Special Education Technology, 24,* 17–30.

Borwein, J. M., & Bailey, D. H. (2005, May). Experimental mathematics: Examples, methods and implications. *Notices of the American Mathematical Society, 52*(5), 502–514.

Boylan, H. R. (2002). What works: Research-based best practices in developmental education. Boone, NC: Continuous Quality Improvement Network with the National Center for Developmental Education. *Education, 108,* 87–101.

Brittenham, R., Cook, R., Hall, J. B., Moore-Whitesell, P., Ruhl-Smith, C., Shafii-Mousavi, M., et al. (2003). Connections: An integrated community of learners. *Journal of Developmental Education, 27*(1), 18–25 (Fall).

Chen, N. S., Teng, D. C. E., Lee, C. H. Kinshuk. (2011). Augmenting paper-based reading activity with direct access to digital materials and scaffolded questioning. *Computers & Education, 57,* 1705–1715. https://doi.org/10.1016/j.compedu.2011.03.013.

Cotton, K. (2001). *Classroom questioning. The schooling practices that matter most.* Northwest Regional Educational Laboratory.

D'Agostino, B. D., Belanger, A., & D'Agostino, B. D., Jr. (1990, November). A suggestion for using powerful and informative tests of normality. *The American Statistician, 44*(4), 316–321.

Dewey, J. (1958). The experimental theory of knowledge. In J. A. Boydston (Ed.), *John Dewey, the middle works* (Vol. 3). Carbondale: University of Southern Illinois Press.

Engelbrecht, J., & Harding, A. (2005). Teaching undergraduate mathematics on the internet. *Educational Studies in Mathematics, 58*(2), 235–252.

Hannafin, R. D., & Foshay, W. R. (2008, April). Computer-based instruction's (CBI) rediscovered role in K–12: An evaluation case study of one high school's use of CBI to improve pass rates on high-stakes tests. *Educational Technology Research & Development, 56*(2), 147–160.

International Technology Education Association. (2006). *Advancing excellence in technological literacy: Student assessment, professional development, and program standards.* Reston, VA: Author.

Jacobson, E. (2006). Computer homework effectiveness in developmental mathematics. *Journal of Developmental Education, 29*(3), 2–8 (Spring).

Kilpatrick, J., Swafford, J., & Findell, B. (2001). *Adding it up: Helping children learn mathematics.* Washington, DC: National Academies Press.

Kim, J. (2011). Developing an instrument to measure social presence in distance higher education. *British Journal of Educational Technology, 42,* 763–777.

Kinney, D. P., & Robertson, D. F. (2003). Technology makes possible new models for delivering developmental mathematics instruction. *Mathematics and Computer Education, 37*(3), 315–328.

Kulik, C.-L. C., & Kulik, J. A. (1991). Effectiveness of computer-based instruction: An updated analysis. *Computers in Human Behavior, 7*(1), 75–94.

Lerman, S. (2001). Cultural, discursive psychology: A sociocultural approach to studying the teaching and learning of mathematics. *Educational Studies in Mathematics, 46,* 87–113.

Lesh, R., Zawojewski, J., & Carmona, G. (2003). *What mathematical abilities are needed for success beyond school in a technology-based age of information?* Mahwah, NJ: Lawrence Erlbaum Associates Inc.

Lester, F. K., & Kehle, P. E. (2003). From problem solving to modeling: The evolution of thinking about research on complex mathematical activity. In R. A. Lesh & H. M. Doerr (Eds.), *Beyond constructivism: Models and modeling perspectives on mathematics problem solving, learning, and teaching* (pp. 501–518). Mahwah, NJ: Lawrence Erlbaum Associates.

Magiera, M. T., & Zawojewski, J. (2011). Characterizations of social-based and self-based contexts associated with students' awareness, evaluation, and regulation of their thinking during small-group mathematical modeling. *Journal for Research in Mathematics Education, 42*(5), 486–520.

Maccini, P., & Gagnon, J. C. (2002). Perceptions and application of NCTM standards by special and general education teachers. *Exceptional Children, 68,* 325–344.

Maccini, P., Mulcahy, C. A., & Wilson, M. G. (2007). A follow-up of mathematics interventions for secondary students with learning disabilities. *Learning Disabilities Research and Practice, 22*(1), 58–74.

Muis, K. R. (2004). Personal epistemology and mathematics: A critical review and synthesis of research. *Review of Educational Research, 74,* 317–377.

National Assessment of Educational Progress (NAEP). (2006). *Writing objectives 2006-07 assessment.* Princeton, NJ: National Assessment of Educational Progress, Educational Testing Service.

National Center for Educational Statistics [NCES]. (2003). Remedial education at degree-granting postsecondary institutions in fall 2000. Retrieved October 4, 2007, from http://nces.ed.gov/surveys/peqis/publications/2004010/.

National Center for Educational Statistics [NCES]. (2005). Trends in educational equity of girls and women: 2004. Retrieved December 17, 2008, from http://www.nces.ed.gov/pubs2005/equity/

National Council of Teacher of Mathematics. (2000). *Principles and standards for school mathematics.* Reston, VA: Author.

National Mathematics Advisory Panel. (2008). *Foundations for success: The final report of the National Mathematics Advisory Panel.* Washington, DC: U.S. Department of Education.

Pearson Education, Inc. (2005). *MyMathLab student results.* Retrieved June 1, 2012, from www.mymathlab.com/success_passrates.html.

Perez, S., & Foshay, R. (2002). Adding up the distance: Can developmental studies work in a distance learning environment? *T.H.E. Journal, 29*(2), 19–24.

Pew Internet and American Life Project. (2002, September 15). *The internet goes to college.* Retrieved October 29, 2008, from http://www.pewinternet.org/pdfs/PIP_College_Report.pdf.

Piaget, J., & Garcia, R. (1991). *Toward a logic of meanings.* Hove and London: Erlbaum (original work published 1987).

Pólya, G. (1957). *How to solve it: A new aspect of mathematical method.* Garden City, NY: Doubleday.

Santos-Trigo, M. (2007). Mathematical problem solving: An evolving research and practice domain. *ZDM, 39,* 523–536.

Schoenfeld, A. H. (2007). Problem solving in the United States, 1970–2008: Research and theory, practice and politics. *ZDM Mathematics Education, 39,* 537–551.

Sfard, A. (2006). Participationist discourse on mathematics learning. In Maasz & Schloeglmann (Eds.), *New mathematics education research and practice* (pp. 153–170).

Smith, G. G., & Ferguson, D. (2004, September/October). Diagrams and math notation in e-learning: Growing pains of a new generation. *International Journal of Mathematical Education in Science and Technology, 35*(5), 681–695.

Stinson, K., Harkness, S. S., Meyer, H., & Stallworth, J. (2009). Mathematics and science integration: Models and characterizations. *School Science and Mathematics, 109*(3), 153–161.

Testone, S. (1999). On-line courses: A comparison of two vastly different experiences. *Research & Teaching in Developmental Education, 16*(1), 93–97 (Fall).

Trenholm, S. (2006). A study on the efficacy of computer-mediated developmental math instruction for traditional community college students. *Research & Teaching in Developmental Education, 22*(2), 51–62 (Spring).

Vygotsky, L. (1978). *Interaction between learning and development. From: Mind and society* (pp. 79–91). Cambridge, MA: Harvard University Press.

Wadsworth, J. H., Husman, J., Duggan, M. A., & Pennington, M. N. (2007). Online mathematics achievement: Effects of learning strategies and self-efficacy. *Journal of Developmental Education, 30*(2), 6–14 (Spring).

Woodward, J. (2004). Mathematics education in the United States: Past to present. *Journal of Learning Disabilities, 37,* 16–31.

Zbiek, R. M., Heid, M. K., Blume, G. W., & Dick, T. P. (2007). Research on technology in mathematics education. In F. K. Lester Jr. (Ed.), *Second handbook of research on mathematics teaching and learning* (pp. 1169–1207). Charlotte, NC: Information Age Publishing.

Chapter 5
Lessons Learned from a Calculus E-Learning System for First-Year University Students with Diverse Mathematics Backgrounds

Lixing Liang, Karfu Yeung, Rachel Ka Wai Lui, William Man Yin Cheung and Kwok Fai Lam

Abstract First-year science majors at The University of Hong Kong have different levels of proficiency in mathematics, with a significant proportion lacking the necessary calculus background for a compulsory freshman science foundation course. A supplementary calculus e-learning platform was implemented so that students lacking the prerequisite could gain the necessary knowledge and skills at their own pace. This chapter presents quantitative and qualitative analyses of the learning analytics, including the behavior as well as the achievements of the users. Pretest and posttest results are used to assess the effectiveness of the platform. Questionnaires completed by the users are utilized to explore aspects for improvement. We hope this study can stimulate discussions on the assessment of e-learning, as well as shed light on the factors contributing to the efficiency and effectiveness of similar platforms.

Keywords Effectiveness of e-learning · Learning analytics · Online calculus platform · Diverse student background

L. Liang · K. Yeung · R. K. W. Lui (✉) · W. M. Y. Cheung · K. F. Lam
Faculty of Science, G/F, Chong Yuet Ming Physics Building,
The University of Hong Kong, Pokfulam Road, Pokfulam, Hong Kong
e-mail: lui2012@hku.hk

L. Liang
e-mail: leoliang@hku.hk

K. Yeung
e-mail: h1352674@hku.hk

W. M. Y. Cheung
e-mail: willmyc@hku.hk

K. F. Lam
e-mail: hrntlkf@hku.hk

© Springer International Publishing AG, part of Springer Nature 2018
J. Silverman and V. Hoyos (eds.), *Distance Learning, E-Learning and Blended Learning in Mathematics Education*, ICME-13 Monographs,
https://doi.org/10.1007/978-3-319-90790-1_5

5.1   Introduction

SCNC1111 *Scientific Method and Reasoning* is a compulsory course for all undergraduate students who intend to pursue a major offered by the Faculty of Science at The University of Hong Kong. It aims to give students a holistic view of the nature and methodology of science, to equip students with basic skills of logical and quantitative reasoning, and to introduce to students a wide spectrum of mathematical and statistical methods for studies and research. Since September 2012, this course has been offered in every fall and spring semester, with enrollment number constantly beyond 200 per semester. For instance, 256 students completed both the midterm and final examinations in the spring semester of 2015–16, on which our research is based.

This course receives students from a variety of educational backgrounds, including students who have taken the Hong Kong Diploma of Secondary Education (HKDSE), the International Baccalaureate (IB) curriculum, the National Joint College Entrance Examination (Chinese NJCEE or Gaokao), the General Certificate of Secondary Education (GCSE) or the Scholastic Assessment Test (SAT). A major challenge in teaching this course is the disparity of the levels of mathematics preparation among such a diverse group of students. Prior to taking the course, a significant proportion of students in the class have not learned any calculus, while the others have different levels of mastery, ranging from being able to apply some basic rules to possessing a comprehensive skillset for advanced computation. As calculus is a major mathematical tool for advanced science learning, equipping all our students with a common level of it has been an essential and critical goal of SCNC1111. In the past, additional tutorials were held for students less prepared in calculus. Conducting these additional tutorials, however, was not very effective since not every student could attend due to scheduling conflicts, and students who managed to attend still had different mathematics background. As a result, the diverse individual learning needs continue to be an issue to be addressed during these tutorials.

To tackle such problem, an e-learning platform was developed in 2015 with which students can learn calculus online through instructional videos that cover topics of functions, limits, differentiation, integration and ordinary differential equations. Students can also use this system to work on quizzes and exercises as well as receive instant feedback. A very powerful feature of such an e-learning platform is that all activities are documented, making it possible to track and analyze students' learning.

The remainder of this chapter is organized as follows. Section 5.2 reviews relevant literature on e-learning. Section 5.3 describes the design of our platform, while Sect. 5.4 outlines the methodology framework employed in this chapter. Data analysis is conducted in Sect. 5.5 to investigate what kinds of students used the platform, how users used it, and what they achieved with it. Section 5.6 offers policy recommendations for improving the e-learning platform with the assistance of a survey conducted by the SCNC1111 Teaching Team. We will finally conclude in Sect. 5.7.

5.2 Literature Review

E-learning systems provide students with non-linear access to information presented in a wide range of formats, including text, graphics and videos (Daradoumis, Bassi, Xhafa, & Caballé, 2013; Jacobson & Archodidou, 2000). It has become a promising alternative to traditional classroom learning. E-learning allows students to learn at their own pace and is able to provide frequent assessment as well as feedback to help students monitor their own learning (Gibbs, 2014; Zhang & Nunamaker, 2003). E-learning systems also contain tools for mass grading, including machine grading and peer assessment, which allow the course to scale to a large number of learners (Daradoumis et al., 2013). E-Learning can be as effective as traditional instruction methods, and several studies have shown that e-learners demonstrate increased content retention and higher student engagement (Chen, Lambert, & Guidry, 2010; Clark, 2002; Nelson Laird & Kuh, 2005; Robinson & Hullinger, 2008).

E-learning systems for mathematics in universities can improve learning outcomes. In a basic calculus course at the University of Helsinki, students taught through an e-learning system had a lower dropout rate than students taught under the traditional instruction method. They also performed better than the traditional students in the posttest (Seppälä, Caprotti, & Xambó, 2006). E-learning systems have also been welcomed by students. In student feedback for an online linear algebra course at Pompeu Fabra University, most students indicated that they were satisfied with the course and had found the materials useful for reviewing concepts as well as for their other courses (Daza, Makriyannis, & Rovira Riera, 2013). The use of online assignments at Simon Fraser University for several calculus courses over a period of five years had also received positive feedback from students. Survey data from the involved students indicate that online assignments led them to read the textbook and lecture notes on a regular basis, and had helped them to better learn the material (Jungic, Kent, & Menz, 2012). E-learning platform has also been devised for use by engineering students at the Universities of Salerno and of Piemonte to generate personalized online courses based on the student's needs (Albano, 2011).

In addition, the use of tests and exercises in e-learning systems has been shown to be effective at the university and secondary levels (De Witte, Haelermans, & Rogge, 2015; Galligan & Hobohm, 2015). Self-testing produces positive effects on student learning by encouraging repeated retrieval of the content being tested, helping students to learn from the feedback they receive, and directing students' further study toward the material that they have yet to master (Black & Wiliam, 1998; Karpicke & Roediger, 2008).

5.3   Platform Design

As presented in Table 5.1, our calculus e-learning platform is divided into five modules: Functions, Limit and Introduction to Differentiation, Differentiation Rules, Application of Differentiation, and Integration. Each module contains three features: videos, quizzes and exercises. Combined with other learning activities of the course, our platform follows the Conceptualization Cycle proposed by Mayes and Fowler (1999) as its pedagogical framework, namely following a three-step approach of conceptualization, construction and dialogue.

Conceptualization, where students acquire information from the videos on the platform, is the first step of the Cycle. The videos present materials in a form similar to a slideshow with voice-over, and most of them last 3–6 minutes. They can be classified into two categories: introduction to theories and discussion of examples. The former, which accounts for 32 of the total 46 videos, gives students the most basic understanding of certain concepts. The illustration is typically carried out in a three-step approach. Firstly, mathematical definitions are proposed, sometimes after discussing the necessary background. For instance, the logistic function is defined after introducing the need to model population growth in a confined environment by biologists. Secondly some of the most basic properties and/or techniques related to this concept are introduced, such as the use of first derivative test to distinguish a local minimum from a local maximum. Sometimes, the illustration is aided with graphs, for example showing the relationship between instantaneous rate and averaged rate in the discussion of differentiation. Thirdly, what has been introduced is applied to one or two straightforward examples which only require direct application of the concepts. These examples are mainly mathematical and computational, so as to allow the viewers to focus mostly on the concepts they just learned. The rest of the 14 videos are application examples aimed to consolidate students' understanding. For instance, radiometric dating is used to illustrate the application of logarithmic and exponential functions. By watching all the videos sequentially, a student with no calculus background is expected to be able to learn every concept required by the course syllabus and tested in the formal assessments. Students can also choose to view particular videos as they see fit.

The second step of the Cycle, construction, where students apply the concepts to meaningful tasks, is carried out in the quiz and exercise components of the platform. The quizzes are listed below the videos in each module on the homepage of the platform. Each of them has 7–17 questions, mostly (193 out of the total 205 questions) in the form of multiple choice and short answer (10 out of 205). Regardless of the form of the question, the difficulty level of almost every quiz is reduced to its minimum and similar to that of the questions discussed in the videos, i.e. direct application of the concepts. Hence students should be able to finish them with ease after watching all the relevant videos. The platform also offers 8 more challenging exercises. Students' participation in them was low (the average number of submitted attempts being 9.0 compared to 36.7 for the quizzes); the exercises are thus excluded from our analysis so as to avoid bias resulted from a small sample.

Table 5.1 Knowledge point involved in the pretest, the examinations and the e-learning platform

Module	Videos (46 total)	Quizzes (17 total)	Pretest (02/05/2016) (full mark: 12)	Midterm (03/15/2016) (full mark: 4)	Final (05/16/2016) (full mark: 8)
Functions	1. Introduction 2. Polynomial functions 3. Exponential functions 4. Logarithmic functions 5. [Example] exp/log functions: caffeine elimination 6. [Example] exp/log functions: radiometric dating 7. [Example] Logistic functions 8. Trigonometric functions 9. Compositions of functions 10. Difference quotients	– Quiz 1 *(1–10)* – Quiz 2 *(1–10)*	– 1 *(8)* – 2 *(2)*		– 1 *(2)* – 2 *(2)* – 5 *(3, 4)*
Limit and introduction to differentiation	1. Introduction to limit 2. Limit at an undefined point 3. Limit at infinity 4. Introduction to differentiation 5. First principles	– Quiz 3 *(1–5)*	– 4 *(1, 3)* – 5 *(1, 2)*		
Differentiation rules	1. Introduction 2. Power rule 3. Constant multiple rule 4. Sum rule 5. Rules for special functions 6. Quotient rule 7. Chair rule 8. Summary	– Quiz 3 *(1–8)* – Quiz 4 *(1, 2, 3)* – Quiz 5 *(2, 3, 4)* – Quiz 6 *(2, 3, 4, 5, 6, 7)* – Quiz 7 *(3, 5, 6, 7)*	– 6 *(1, 3, 4)* – 7 *(1, 2, 3, 4)* – 8 *(1, 5)* – 9 *(5, 7)* – 10 *(5, 6)*	– 1 *(2, 3, 4, 6)* – 8 *(2, 3, 4, 6)*	– 3 *(5, 6)* – 4 *(5, 6)*

(continued)

Table 5.1 (continued)

Module	Videos (46 total)	Quizzes (17 total)	Pretest (02/05/2016) (full mark: 12)	Midterm (03/15/2016) (full mark: 4)	Final (05/16/2016) (full mark: 8)
Application of differentiation	1. Introduction 2. [Example] Rate of change: second order reaction 3. [Example] Rate of change: koala population 4. Related rates introduction 5. Interpretation of the first derivative 6. First derivative test 7. Interpretation of the second derivative 8. Second derivative test 9. [Example] Optimization: introduction 10. [Example] Optimization: building design 11. Concept of inflection points	– Quiz 8 (5, 7) – Quiz 9 (5, 6, 7, 8, 9) – Quiz 10 (7, 8, 11)		– 2 (5, 7) – 8 (5, 6)	– 4 (5, 6)
Integration	1. Antiderivative 2. [EXAMPLE] Antiderivatives: falling coconut 3. Integration by substitution 4. Definite integral (I) 5. Definite integral (II) 6. Fundamental theorem of calculus 7. [EXAMPLE] FTC: area under a curve	– Quiz 11 (1, 3, 4, 5, 6) – Quiz 12 (1, 2, 3, 4, 5, 6) – Quiz 13 (4, 5) – Quiz 14 (1, 4, 5, 6) – Quiz 15 (1, 4, 5, 6) – Quiz 16 (1, 4, 5, 6) – Quiz 17 (1, 6)	– 11 (1) – 12 (1, 4, 5, 6) – 13 (1, 2, 3, 4, 5, 6)	– 3 (1, 3, 6)	– 3 (1, 3, 6)

(continued)

Table 5.1 (continued)

Module	Videos (46 total)	Quizzes (17 total)	Pretest (02/05/2016) (full mark: 12)	Midterm (03/15/ 2016) (full mark: 4)	Final (05/16/2016) (full mark: 8)
	8. [EXAMPLE] FTC: area between 2 curves 9. [EXAMPLE] FTC: population density 10. [EXAMPLE] Modelling using DE: learning curve 11. [EXAMPLE] DE: population growth 12. [EXAMPLE] DE: lake pollution				

The numbers in the brackets behind each question refer to the videos that cover the necessary calculus knowledge for its standard solution, as provided by the SCNC1111 Teaching Team. For instance, the "(1–10)" in the bracket behind Quiz 1 means that in order to solve all questions in Quiz 1 using the standard solution, all one needs to know is the content introduced in Video 1-10 in Module 1

The third step of the Cycle, dialogue, where students engage in meaningful conversations and discussions with their tutors and peers, is carried out both online and offline within the course. An open discussion forum is established on the SCNC1111 course website, where both instructors and students participated. In addition, weekly face-to-face small-group tutorial sessions were held, offering opportunities to discuss course materials.

5.4 Methodology

Our e-learning platform was offered to all the students of SCNC1111, and the data analyzed in this chapter are from the spring semester of 2015–16. Through this analysis, we seek to answer the following questions:

1. Do students with different calculus background use our platform with different intensity?
2. Are different features of the platform (i.e. the videos and the quizzes) utilized differently by the students?
3. Is there quantitative evidence to show an association between student improvement in calculus knowledge and the use of the platform?

The user log of the e-learning platform, where a student is the basic unit of observation, is the starting point of our quantitative analysis. We aggregate the user log into multiple variables that describe each user's intensity of usage. To assess the students' prior calculus knowledge, a pretest, which only allows one submission, was carried out on the SCNC1111 official course website. Students had been reminded multiple times in tutorials, lectures and through emails to finish the pretest, and they were given 3 weeks to complete it independently without prior revision. To encourage them to show their actual ability without seeking external assistance, students were assured that the pretest was not linked to their final grades. Twelve of the pretest questions require only direct applications of theorems, with difficulty level similar to that of the platform quizzes. We then differentiate different types of students based on their response to the pretest.

To answer Question 1 above, we conduct two-sample t-tests and linear regressions to analyze the correlation between variables on students' intensity of platform usage and variables on students' knowledge background. These variables regarding user's intensity of participation also allow us to answer Question 2 via applying linear regression models to explore their association with the characteristics of the user.

To answer Question 3, measures are needed to indicate the student's performance in the course, and we have picked the calculus part of the midterm and final examinations for this purpose. The midterm examination on March 15, 2016 contains 4 questions related to calculus, with levels of difficulty similar to those of the quizzes. The final examination on May 16, 2016 requires more steps to solve each question (an integration of the knowledge rather than simple direct application). In addition, the form of assessment was changed to long-answer questions in which students were required to state clearly every step of their reasoning. How the calculus knowledge is aligned between the platform and other assessments of the course is displayed in Table 5.1.

With these data on platform usage and performance in course assessments, t-test and linear regressions are then conducted to investigate their inter-relationship. It is noteworthy that the data available to this research constrained us from making statements about causal relationships, and hence our analysis is mostly about observations on correlation. Proposals to record and collect data for better inference will be discussed after key findings are presented.

To overcome the limitations of our data and understand better students' motivations as well as needs so as to improve the e-learning platform, a survey, partly inspired by the research of Sun, Tsai, Finger, Chen, and Yeh (2008), was designed to assess the platform comprehensively. Students' responses are expected to help us better understand the more subtle details that our quantitative analysis may fail to capture. Due to space limit, we are only able to analyze parts of the survey responses. The discussion will be focused on the major questions and observations found in the data analysis in Sect. 5.5.

5.5 Data Analysis

This section is organized as follows. Section 5.5.1 defines various measures of student participation in the platform and their learning achievement. Section 5.5.2 analyzes the relationship between the intensity of platform participation and students' prior calculus knowledge. Section 5.5.3 discusses how platform users with different prior knowledge utilized videos and quizzes on the platform. Section 5.5.4 explores the correlation between student improvement and platform participation. Section 5.5.5 summarizes the key findings, and Sect. 5.5.6 suggests aspects for improvement in the research.

5.5.1 Data Collection and Descriptive Statistics

Among the total 256 students that took both the midterm and final examinations, 225 submitted the pretest and form our sample. To differentiate the student population based on their background in calculus, we can use either their scores in the pretest or their self-assessment. In the pretest, students were asked the question "Have you learned calculus before?" The variable *rBackground*, which forms our measure for students' self-assessment, is defined as 1 if they identify themselves as having at least some prior knowledge of calculus, and 0 otherwise. As our first step, we would like to ensure such self-assessment and the pretest score perform equally

Table 5.2 Pretest scores for students who self-reported as having and not having prior knowledge in calculus

Variable	Self-reported prior knowledge = 0 (n = 71)	Self-reported prior knowledge = 1 (n = 154)	t-test
	Mean (sd)	Mean (sd)	t-statistic (*p*-value)
Pretest score (out of 100)	25.94 (25.14)	84.74 (17.56)	$t(102.67) = -17.806$ (<0.001)

One-sided two-sample t-test allowing for unequal variance is used to evaluate whether the pretest score is lower for students who reported they do not have prior knowledge in calculus

well in representing the students' prior calculus knowledge. Here we standardize the pretest score such that the maximum score achievable is 100, so that we can interpret the score as the percentage of marks obtained. While it is true that students' self-assessment is not absolutely accurate, the t-test[1] in Table 5.2 shows that it is able to differentiate the two groups of students whose pretest scores are significantly different. On average, students who reported themselves as not having prior calculus knowledge gained 25.94% of the score in the pretest, while the other group gained 84.74%. The pretest score is significantly lower for students (self-reported as) without calculus background at the 1% significance level. This means *rBackground* and the pretest score can be equally good indicators for our coming analyses to represent students' prior calculus background.

We then define the following variables as measures of participation in the platform. Firstly, for usage in videos, *qvideo* measures the total number of videos watched,[2] while *cvideo* makes a similar measurement except that repeated watching of the same video by the same student will only be counted once.[3] Similarly, to capture how often the quizzes are utilized by the students, *qquiz_all* and *cquiz_all* measure the number of quizzes submitted (with or without double-counting repeated submission of the same quiz by the same student). The number of quizzes submitted, however, cannot capture the quality of submission. For example, a full mark in a quiz may reveal different learning outcome or effort than a zero score. To address this, we define the variables *qquiz_pass* and *cquiz_pass*. The definitions of these two are similar to the previous ones, except that any quiz attempts that scored less than 30% of the full mark are excluded. The descriptive statistics for all these variables are listed in Table 5.3.

---

[1]The t-test allowing for unequal variance is used as an F-test comparing the variances of pretest scores of the two groups suggests that we cannot treat the variances as the same at 1% significance level.

[2]Due to limitations in the user log, we define "watching a video" as clicking the video link. The assumption is that users watch the video once every time they click the link. Unfortunately, this cannot capture the case where a user stays on the same page and watch a video over and over again.

[3]For example, if a user had clicked on the link of one particular video n times throughout the semester, his/her *qvideo* would be n while *cvideo* would only be 1.

Table 5.3 Descriptive statistics

Variable	No. of observations	Mean	St. Dev.	Min	Max
Assessment results					
Pretest (full mark = 12)	225	7.9	4.1	0	12
Midterm (full mark = 4)	225	2.6	1.1	0	4
Final (full mark = 8)	225	4.0	2.0	0	8.0
Platform usage					
Videos watched (qvideo)	225	17.6	27.7	0	161
Unique videos watched (cvideo)	225	10.9	15.1	0	46
Quizzes submitted (qquiz_all)	225	2.5	4.5	0	36
Unique quizzes submitted (cquiz_all)	225	2.1	3.4	0	17
Quizzes passed (qquiz_pass)	225	2.1	3.8	0	25
Unique quizzes passed (cquiz_pass)	225	1.9	3.1	0	17

Tables 5.3, 5.4 and 5.5 were generated using the R-package Stargazer (Hlavac, 2015)

5.5.2 Relationship Between E-Learning Participation and Calculus Background

Understanding how different groups of students used an e-learning platform can inform us how to develop one that suits the needs of each group better. Our e-learning platform can potentially offer us such understanding because it was designed to be a supplementary rather than principal part of the SCNC1111 curriculum. Participation is optional but welcomed regardless of students' prior knowledge background. Thus, our platform was used by different types of students for multiple purposes (e.g. study, practice and revision) depending on their needs and preferences. We start our analysis by defining the variable *use* for each student to differentiate those who have not accessed the platform at all: *use* being 1 if the student has either watched one video or attempted one quiz, and 0 otherwise.

Regression was performed to investigate how students' "platform participation" variables are related to their calculus background (as indicated by their pretest score). The results are presented in Table 5.4 column (1) treats whether the student has used the platform as the dependent variable, while columns (2) and (3) use the number of unique videos watched as the response. In column (1), we observe that the higher the score on pretest is, the less likely one will use the platform ($\beta = -.04$, $p < .001$). In column (2), we can observe that a higher pretest score is associated with fewer number of videos watched ($\beta = -1.68$, $p < .001$). In column (3), we limit the analysis to cases where the user has accessed the platform (i.e. *use* = 1). We can see that the same relationship between number of videos watched and calculus background is still significant ($\beta = -1.59$, $p < .001$) among users who have at least watched one video or submitted a quiz.

Table 5.4 Relationship between e-learning participation and calculus background

	Dependent variable						
	Use	Unique videos watched (cvideo)		Unique quizzes submitted (cquiz_all)		qvideo/cvideo	qquiz_all/cquiz_all
	(1)	(2)	(3)	(4)	(5)	(6)	(7)
Pretest	−0.036***	−1.675***	−1.587***	−0.115**	0.017	−0.033***	0.010
	(0.008)	(0.221)	(0.292)	(0.055)	(0.077)	(0.009)	(0.006)
Constant	0.883***	24.188***	29.184***	3.046***	3.429***	1.607***	1.046***
	(0.069)	(1.977)	(2.379)	(0.493)	(0.628)	(0.068)	(0.049)
Limit to use = 1			Yes		Yes	Yes	Yes
Limit to qvideo > 0						Yes	
Limit to qquiz_all > 0							Yes
Observations removed	0	0	0	0	0	2	2
Observations remain	225	225	135	225	135	126	96
R^2	0.088	0.204	0.182	0.019	0.0004	0.106	0.027
Adjusted R^2	0.084	0.200	0.175	0.015	−0.007	0.098	0.016
F Statistic	21.501*** (df = 1; 223)	57.167*** (df = 1; 223)	29.493*** (df = 1; 133)	4.366** (df = 1; 223)	0.049 (df = 1; 133)	14.635*** (df = 1; 124)	2.588 (df = 1; 94)

As displayed in columns (6) and (7), some observations are removed from the regression model. Those removed are influential (i.e. high leverage) observations and/or outliers. Throughout this chapter, an observation will be removed if and only if its Cook's Distance is larger than the 25th percentile of the F distribution with degrees of freedom the same as that of the regression model. In addition, in this table, the signs and significance of the coefficients do not change when those observations are not dropped in the first place

At the significance level of 5%, in columns (2) and (3), the significance and signs of the coefficients do not change if the dependent variable is *qvideo* rather than *cvideo*. Similarly, in columns (4) and (5), our conclusion does not change if the dependent variable is *qquiz_all* rather than *cquiz_all*

In columns (4) and (5), we have also run regressions with *qquiz_pass* or *cquiz_pass* as the response. At the 5% level, the association between either of these variables and the pretest score is not significant

An entry of "Yes" in the row "Limit to use = 1" indicates that our regression analysis for that column is limited to cases with the variable *use* = 1. Rows with similar wording convey similar messages

Note *$p < 0.1$; **$p < 0.05$; ***$p < 0.01$

Considering that quizzes are the other major part of the e-learning platform, columns (4) and (5) use the number of quizzes submitted as the dependent variable. They have similar settings as those in columns (2) and (3) respectively. In column (4), we find that the number of quizzes submitted is negatively associated with a student's prior calculus background ($\beta = -.13$, $p = .04$). However, when we limit our sample to those who have at least used one feature of the platform once in column (5), such association is no longer significant, even at the 10% significance level. This means that among genuine platform users, the relationship between the intensity of submitting quizzes and prior knowledge background is not significant. The relationship observed in column (4) might be just due to a number of students with stronger prior knowledge background who are much less likely to use the platform, as indicated in column (1).

Columns (6) and (7) discuss the intensity of platform usage from a different angle. In addition to the number of videos or quizzes watched/submitted, we can analyze how many times on average a student accesses the same platform item. Such average frequency[4] can be represented by *qvideo/cvideo* (for watching a video repeatedly) and *qquiz_all/cquiz_all* (for submitting a quiz more than once). Their relationships with the pretest score are shown in columns (6) and (7). We observe in column (6) that the average number of views per video across time is negatively associated with the user's pretest score ($\beta = -.03$, $p < .001$). Similar relationship is shown for the quizzes in column (7), although it is not significant even at the 10% level ($p = .111$).

Based on the above findings, we can draw three conclusions regarding the relationship between e-learning participation and students' calculus background. Firstly, students with a stronger background in calculus (as measured by their pretest score[5]) are less likely to use the platform. Secondly, students with a weaker prior background in calculus tend to watch more videos, both in the sense that they watch a larger number of different videos (i.e. higher *cvideo*) and watch the same video repetitively for more times (i.e. higher *qvideo/cvideo*). Thirdly, there is no significant association between attempts in quizzes and student's calculus background. In other words, we cannot conclude that students with different calculus background use the quiz function of our platform with different intensity, in terms of attempting different quizzes and making repetitive attempts of the same quiz. More intuitively speaking, students with weaker prior knowledge tend to watch

---

[4]Average frequency refers to the number of times a student watched the same video or submitted the same quiz. For instance, if a student has watched 10 different videos (*cvideo* = 10) with a total view count of 30 (*qvideo* = 30), then his/her average frequency of watching videos is defined as *qvideo/cvideo* = 3.

[5]As for column (1) of Table 5.4, the same conclusion can still be reached if the independent variable is students' self-reported calculus background (*rBackground*) instead of their pretest score.

videos on more different topics as well as re-watch the videos for more times, yet they are not as eager to participate in quiz practices. This situation will be further discussed in the following sections.

5.5.3 Relationship Between Features of the Platform

The previous section discusses in what intensity students with different calculus backgrounds use the different functions on the e-learning platform. In this section, we use similar regression techniques to capture the relationship between the two major features of our platform, the quizzes and the videos. Throughout this section, we limit the sample to students who have at least watched one video or submitted one quiz (*use* = 1).

There are two main questions to be addressed. Firstly, we wonder if students who watch the videos intensively tend to attempt the quizzes intensively as well. In other words, we analyze if the quizzes and the videos are substitutes for or complements to each other. Secondly, we want to know, given a certain number of videos viewed, whether students with different calculus backgrounds tend to attempt quizzes with different intensity.[6]

The results are presented in columns (8) and (9) of Table 5.5. Column (8) treats the number of unique quizzes submitted (*cquiz_all*) as the response variable, while column (9) considers the number of unique quizzes passed (*cquiz_pass*) as the response. In both cases the explanatory variables are the number of unique videos watched (*cvideo*) and the student's score in the pretest.

From columns (8) and (9), we can see a positive and significant association between the intensity of quiz attempts, as measured by the number of unique quizzes submitted or passed, and the intensity of watching videos, as measured by the number of unique videos watched, given the same level of pretest (i.e. positive coefficient for *cvideo*, β = .13 and .11 respectively with $p < .001$ in both cases). In other words, quizzes and videos are complementary to each other for the students' learning via the platform.

In addition, the coefficients for pretest in columns (8) and (9) are also positive and significant (i.e. β = .23 and .25 with $p = .003$ and $p < .001$ respectively). One interpretation is that given the same intensity in watching videos (as measured by *cvideo*), students with stronger prior knowledge in calculus tend to use the quiz feature more intensively. Considering that this platform was originally designed for students with weaker prior knowledge, this is contrary to the intuition that students

---

[6]Originally, we also wanted to analyze whether and how students' prior knowledge affects the substitute/complement relationship between the videos and the quizzes. Hence, the interaction effect between *cvideo* and *pretest* was initially included in the full model. At the significance level of 5%, the interaction effect was not statistically significant. In other words, the relationship between quizzes and videos is not affected by a student's calculus knowledge background. In columns (8) and (9), therefore, only the main effects of *cvideo* and *pretest* are included.

Table 5.5 Relationship between features of the platform

	Dependent variable			
	Unique quizzes submitted (cquiz_all)	Unique quizzes passed (cquiz_pass)	Improvement (sdFinal − sdPretest)	Improvement (sdMidterm − sdPretest)
	(8)	(9)	(10)	(11)
cvideo	0.132***	0.109***	0.394**	0.866***
	(0.020)	(0.019)	(0.170)	(0.209)
Pretest	0.226***	0.246***		
	(0.074)	(0.071)		
cquiz_all			−0.949	−0.972
			(0.712)	(0.862)
Constant	−0.421	−0.593	−15.111***	−7.907*
	(0.797)	(0.762)	(3.840)	(4.771)
Limit to use = 1	Yes	Yes	Yes	Yes
Observations	135	135	135	126
R^2	0.250	0.207	0.040	0.125
Adjusted R^2	0.239	0.195	0.025	0.111
F statistic	22.038*** (df = 2; 132)	17.210*** (df = 2; 132)	2.747* (df = 2; 132)	8.796*** (df = 2; 123)

Only columns (8) and (9) are relevant to the analysis of Sect. 5.5.3. Columns (10) and (11) explore the association between learning outcome and platform participation, to be discussed in Sect. 5.5.4
There are no influential (high leverage) observations or outliers that are dropped in Table 5.5
In column (8), the same conclusion regarding the significance and sign of the coefficient can still be drawn even if the response variable is *qquiz_all*, and/or the independent variable *cvideo* is replaced by *qvideo*. In column (9), at the 5% significance level, the same conclusion regarding the significance and sign of the coefficient can still be drawn even if the response variable is *qquiz_pass*, and/or the independent variable *cvideo* is replaced by *qvideo*
Note $*p < 0.1$; $*p < 0.05$; $***p < 0.01$

with weaker calculus background would use both functions of the platform more intensively. This finding is consistent with those from column (4) and (5) in Table 5.4 regarding the relationship between intensity of quiz attempt and prior calculus knowledge. We may conclude that compared with their peers, students with weaker calculus background are not as active in attempting quizzes as they are in watching videos. To some extent, it reveals that students without prior knowledge tend to prefer learning passively through features like videos, but overlooked the importance of practice. This will be discussed further in Sect. 5.6 when offering policy recommendations.

5.5.4 Relationship Between Student Improvement and E-Learning Participation

We used two proxies for the intensity of using the platform. First, we use the binary variable *use* to differentiate students who have at least watched one video or submitted one quiz on the platform from those who do not. Then, a t-test is conducted to compare the mean *improvement* between these two groups of students. The null hypothesis is that there is no difference in *improvement* between them, and the alternative hypothesis is that the students who have not used the platform have relatively lower *improvement* score. The results of this test are presented in Table 5.6. Firstly, from the pretest to the final examination, the mean score differences are both negative for users and non-users of the platform, which can be interpreted as that the final examination itself is more challenging. However, the score difference is more negative or, in other words, relatively lower for students who had not used the platform ($t(223) = -3.28$, $p < .001$). Hence, we shall reject the null hypothesis and conclude that students who used the platform tend to have larger improvement from the pretest to the final examination. Secondly, we can also analyze the period from the pretest to the midterm examination to test the robustness of our above claim. Students who used the platform showed a positive score difference (mean = 4.29) whereas students who did not use the platform showed a negative score difference (mean = −4.39). Such difference is statistically significant ($t(223) = -1.94$, $p = .03$), and hence we can conclude that students who used the platform between the beginning of the semester and the midterm examination tend to have higher improvement, as observed from differences in test scores. These analyses show that platform users tend to have better improvement in the midterm and final examinations compared to their peers who do not use the platform.

Table 5.6 Relationship between improvement and platform usage

Variable: improvement	Did not used the platform use = 0	Used the platform use = 1	t-test
	mean (sd)	mean (sd)	t-statistic (*p*-value)
sdFinal − sdPretest	−23.76 (27.21)	−11.319 (28.27)	$t(223) = -3.2827$ (<0.001)
sdMidterm − sdPretest	−4.39 (30.64)	4.29 (35.29)	$t(223) = -1.9391$ (0.027)

In both rows, we employ t-tests that assume equal variance because both F-tests comparing the variances of *improvement* of the two groups suggest the variances are the same, at the 1% level
The two rows cover different time spans, so the number of observations in each group differs. From the start of the semester to the midterm examination, 126 out of the 225 students used the platform (i.e. *use* = 1). By the date of the final examination, 135 out of the 225 students have used it
One-sided two-sample t-tests assuming equal variance are used to evaluate whether improvement between posttest and pretest is lower for students who have not used any functions on the platform

The second way to proxy the intensity of platform usage is the number of videos watched and quizzes submitted. In columns (10) and (11) of Table 5.5, we treat the improvement from the pretest to the final and midterm examinations as the response variables respectively. We can observe that there is a positive and significant association between the improvement and the intensity of watching videos, as measured by the number of unique videos watched (i.e. positive coefficient for *cvideo*, with $\beta = .39$ and $.87$, $p = .02$ and $< .001$, in columns (10) and (11) respectively). Both coefficients for *cvideo* are statistically significant at the 5% significance level.

However, there seems not to be a statistically significant association between improvement of scores and intensity of participation in quizzes. The coefficients for *cquiz_all* are not statistically significant even at the 10% significance level in both columns (10) and (11). The interpretation may be as follows. As it is discussed in Sects. 5.5.2 and 5.5.3 (based on observations from columns (2) and (3) in Table 5.4 and from column (8) and (9) in Table 5.5), there is no significant negative association between participation in quizzes and students' prior knowledge background once we control for other variables such as *use*. In other words, students with stronger calculus background are equally as likely to attempt the quizzes as their less well-prepared peers. Therefore, the usage of quizzes is varied, in that students with stronger calculus background may use it for revision and practice, while the others may use it for consolidating newly learned concepts. For the group of students using them for revision, attempting the quizzes may not associate with a higher improvement in scores as there is little room for improvement for students who are already well-prepared in calculus.

5.5.5 Key Findings from Data Analyses

From the data analyses above, we are able to answer the three questions proposed in Sect. 5.4. First, students with weaker prior knowledge in calculus are more likely to use the platform intensively. In particular, they display a preference for videos, as shown in their tendency to watch more videos of different topics and watch the same video repetitively. However, they are not as keen in practicing the quizzes, in the sense that the intensity of quiz participation is not associated with a student's calculus background. The association even seems to be positive once we control for other variables. Second, a complementary relationship between the two functions of the platform, the videos and the quizzes, is present, regardless of the user's prior knowledge. Third, improvement between the examinations and the pretest is positively associated with the usage of our platform. Though our observations cannot be proved as causal, they can still provide insight on the behavior and preference of students with diverse learning needs.

5.5.6 Reflections on Research Improvement

While the above t-tests and linear regression models provide a consistent picture of user behavior and learning outcome, our ability to infer patterns is still limited by multiple elements. Developing a more comprehensive dataset and experimenting with methods to identify causality are two major areas similar projects can improve on in the future.

First, our data analysis can be more rigorous by controlling for various fixed effects, so as to account for any nested structure in the data (De Witte et al., 2015). These effects can be the students' familiarity with e-learning and their learning ability. In future offering of the course, the SCNC1111 Teaching Team can include questions related to student's characteristics (such as their familiarity with e-learning) in the pretest, in addition to the existing mathematical questions. Moreover, scores of assignments that are not directly relevant to calculus might be used as a proxy for students' learning ability which can then be controlled for in our regression model. Still, a more thorough and detailed examination will be needed in deciding what data to collect and what effects to control for in future research.

Second, as stressed in earlier analyses, the data available to us did not enable us to explore causality. To improve, the first and most obvious way is through randomized control experiments as adopted in the research of Papastergiou (2009), Potocki, Ecalle, and Magnan (2013); yet doing so may be infeasible or even inappropriate under some teaching settings (such as not granting part of the class access to certain teaching facilities). The Instrumental Variable (IV) approach as discussed in Angrist and Lavy (2002), De Witte et al. (2015), or Machin, McNally, and Silva (2007) can be a possible way to explore causality using only observational data. Integrating elements of a randomized experiment with an IV approach as discussed by Rouse and Krueger (2004) is also worth exploring in future projects.

5.6 Policy Recommendations

The relationship discussed in Sect. 5.5 between different features of the platform and the preferences of students with different calculus background offers us insights on future improvement of e-learning efficiency and effectiveness. To partly overcome the limitations on our data and to gather direct feedback from the students, we designed a comprehensive questionnaire for the SCNC1111 Teaching Team to distribute to the students after the end of the semester. This section will discuss this questionnaire and what we have learned from it. Section 5.6.1 briefly introduces the structure of the questionnaire. Section 5.6.2 offers policy recommendations on implementing e-learning platforms, based on observations from Sect. 5.5 and students' feedback in the questionnaire.

5.6.1 Questionnaire Design

The questionnaire is modeled after the research of Sun et al. (2008), which suggests six groups of factors influencing user experience: the learner, instructor, course, technology, design, and environmental dimensions. In addition, we have included questions specifically related to our e-learning platform, such as those about student's attitude towards and suggestions for the videos and the quizzes. Since the platform is connected to the general SCNC1111 learning experience, questions regarding how well it linked to other components of the course are also included. Students are, furthermore, welcomed to provide open-ended suggestions for future improvement. A summary of the survey is displayed in Table 5.7 in the Appendix. The space limit of this chapter, however, forbids us from conducting a comprehensive review of the responses, which shall be presented in a separate paper. In this section, we instead focus our analysis mainly on the issues identified in Sect. 5.5 and draw inspirations from the survey responses.

5.6.2 Recommendations for Future Improvement

Section 5.5 provides us with insights on how to improve the platform with regard to the relationship among quizzes, videos, and students' prior knowledge in calculus. Here we make three proposals to improve user satisfaction and education effectiveness, namely adjustment of quiz design, appropriate pace of learning, and facilitation of dialogue.

The first proposal aims at increasing users' participation in the quizzes. As we have discussed in Sect. 5.5, students with weaker prior knowledge in calculus are not as keen in doing practice in quizzes as they are in watching videos. If we posit that videos and quizzes are the two major sources for students to complete the first two steps of the Conceptualization Cycle respectively, low participation in quizzes means a weak construction phase (Mayes & Fowler, 1999). While this might be due to different learning needs and preferences,[7] more options can be incorporated into the quizzes to accommodate such diversity, such that more students will be more willing to attempt them. Two such possible ways to enhance the quizzes are to offer practice questions of more levels of difficulty (so that the students can sense more clearly the quizzes are preparing them for the examinations) and to offer the quizzes in various formats.

According to our survey, 95% of the platform users agree that the quizzes align closely with the content of the videos,[8] yet only 83% of the users agree that the quizzes cover all contents in the examinations. Moreover, students tend to consider

---

[7]For instance, students may have managed to find practice questions from other sources.

[8]Defined as assigning number 3, 4, 5 to the statement "the quizzes aligned closely with the content of the videos" on a Likert Scale of 1–5: 1 means "strongly disagree" and 5 means "strongly agree".

the quizzes to be easy compared with questions in the midterm and final exami-nations.[9] As mentioned in Sect. 5.3, quizzes on the platform are designed to be straightforward (direct application of the involved concepts), whereas more chal-lenging questions are included in the exercises. Students who sought more advanced questions, therefore, may already be able to satisfy their needs from the exercises. The reason for such survey response, it seems, is that the platform has not fully informed the students of the differences between the quizzes and the exercises. In this regard, an e-learning platform should ensure that clear information about all its components is conveyed to its users. Furthermore, more advanced questions of more levels of difficulty (such as those that require similar ways of reasoning as in the examinations and those that are in some ways related to examination questions) can be provided. Doing so may help to boost students' confidence in learning (and thus willingness to work on the quizzes) via fostering a sense of gentle and gradual increase in difficulty level. Some students have also indicated in the open-ended response[10] their preference for printable handouts and quizzes. Such preference may be because our examinations are conducted offline and students find these forms of learning materials more in line with the examination format. Providing such kinds of offline learning opportunities on the platform can help encouraging students who prefer learning this way to attempt the quizzes.

Second, it is apparent that some students skipped the quizzes, reporting that "they were running out of time during revision".[11] To take a closer look, we graph the average number of videos watched and quizzes submitted per student per day respectively,[12] where the red vertical lines indicate the dates of the midterm and final examinations (Figs. 5.1 and 5.2). The traffic for both quizzes and videos soared on the immediate periods before examinations. Some may use the platform as a last-minute study shortcut before the examinations. The resulted lack of time for study will prevent students from combining learning with actual practices. For instance, some students may not even be able to spare 10 minutes to finish a quiz, simply because they were rushing through the materials the day before an exami-nation. Furthermore, as some of the contents introduced in our lectures and tutorials (such as separable differential equation) require basic understanding on calculus, the delay in grasping the necessary calculus concepts will make it even harder for those to catch up with the course progress, resulting in a vicious circle of losing confi-dence in mathematics. Therefore, though participation in the e-learning platform

---

[9]Only 16 and 14% of the students respectively disagree or strongly disagree with the statement that "Compared with the midterm/final examination, the quizzes were too easy".

[10]For instance, in the course dimension part of the survey, an optional question asked "What gave you a better learning experience than the e-learning platform in Calculus learning, if any?".

[11]In the quiz part of the survey, a question asked if there is "any other reasons that stopped/prevented you (the students) from working on the quizzes". A number of respondents claim the quizzes are time-consuming or they do not have enough time for revision.

[12]The average is defined as the total number of videos/quizzes accessed every day divided by the total number of students with certain mathematical background. 71 students reported themselves as having no prior calculus background, and 154 students reported they have.

Fig. 5.1 Average number of videos watched

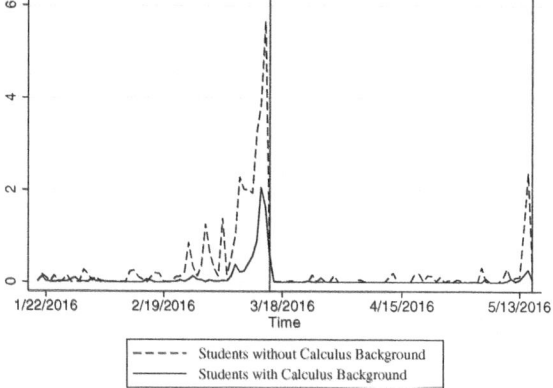

Fig. 5.2 Average number of quizzes submitted

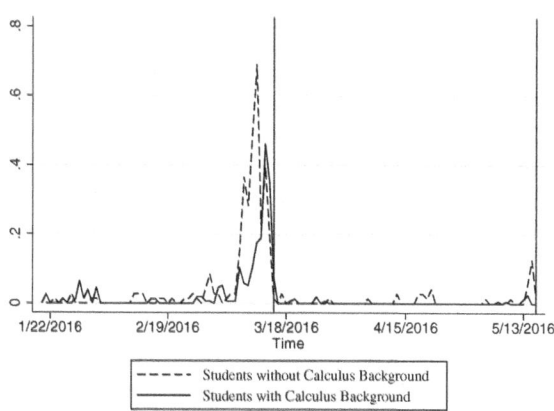

shall still be optional, it would be important to share with the students a recommended timeline for finishing the constituent modules. This would allow the students to plan ahead and allocate enough time for the learning and practice, as well as for better integrating with other related contents of the lectures and the tutorials. In this regard, a number of students have proposed to combine such timeline with printable study notes, such that the learning progress can be more organized.[13]

Last but not least, it would be important for the platform to enhance dialogue among its users. In the Conceptualization Cycle, the third step, dialogue, is crucial as this is when conceptualizations can be tested and further developed. While

---

[13]This is observed in students' response to the question "I would suggest the following changes for improvement:". In addition, 88.5% of the respondents agreed with the proposal that "the instructors should suggest a timeline that we finish certain modules of the e-learning platform".

opportunities for dialogue have already existed in SCNC1111 (such as the all-purpose online forum on the course website and the tutorial sessions with the instructors), the survey still reflects more demand for tools for communication. For instance, some suggest the development of an online Q&A forum within our platform. This can indicate our students' preference for more convenient opportunities to discuss their learning progress with greater focus (i.e. a forum dedicated for discussing calculus learning materials).

5.7 Conclusion

This chapter offers an integrated quantitative and qualitative assessment of the e-learning platform for a compulsory freshman science course, SCNC1111, at The University of Hong Kong. In terms of user behavior, data analysis indicates that users with weaker prior knowledge of calculus tend to utilize the platform, specifically its instructional videos, more intensively compared to their peers with stronger background in calculus. We also observe that the videos and the quizzes of the platform complement each other in the students' learning, regardless of their prior calculus knowledge. In addition, association is identified between greater improvement (from the pretest to the examinations) and a higher intensity of platform participation. While there are limits in data availability, we conclude that participating in the SCNC1111 e-learning platform is positively associated with a better understanding and ability in calculus knowledge. Finally, based on our data analysis and feedback from the platform users, we identify ways to enhance the efficiency and effectiveness of e-learning platforms, namely adjustments to the quizzes, provision of a recommended timeline of learning, and enhanced facilitation of dialogue.

Acknowledgements We would like to acknowledge here the efforts of other members of the SCNC1111 teaching team and the e-learning platform development team who are not on the author list.

Appendix

See Table 5.7.

Table 5.7 Questionnaire items

User background	• Pre-university curriculum • Academic discipline • Previous knowledge of different modules on calculus
Learner dimension	• Students' access to computer and the Internet • Students' familiarity with Information Technologies
Instructor dimension	• Instructors' encouragement on the use of platform • Instructors' commitment to provide timely support
Technology dimension	• Speed and reliability of the connection of the e-learning platform
Course dimension	• Platform's ability to organize the learning process • Platform's ability to make learning easier and more effective
Design dimension	• Perceived ease of use and user-friendly layout
Environmental dimension	• Variety of options for learning • Communication between users and instructors
Videos	• Purpose of watching videos • Whether contents of videos prepare students for examinations • Whether videos present knowledge in an organized manner, at an appropriate pace and at the appropriate level of difficulty
Quizzes	• Purpose of attempting quizzes • Whether quizzes offer accurate assessment of students' learning • Whether quiz practices prepare students better for examinations • Whether quizzes are designed at an appropriate amount and at the appropriate level of difficulty
Midterm examination	• How likely one will skip calculus questions in the midterm examination • Whether the platform prepares students better for the midterm examination
Final examination	• How likely one will skip calculus questions in the final examination • Whether the platform prepares students better for the final examination
Lectures and tutorials	• Whether offline and online learning complement with each other
Future improvement suggestions	• Reasons why some students did not use the e-learning platform • Open-ended proposals for future improvement

References

Albano, G. (2011). Mathematics education: Teaching and learning opportunities in blended learning. In A. Juan, A. Huertas, S. Trenholm, & C. Steegmann (Eds.), *Teaching mathematics online: Emergent technologies and methodologies* (pp. 60–89). Hershey, PA: Information Science Reference.

Angrist, J., & Lavy, V. (2002). New evidence on classroom computers and pupil learning. *The Economic Journal, 112*(482), 735–765. https://doi.org/10.1111/1468-0297.00068.

Black, P., & Wiliam, D. (1998). Assessment and classroom learning. *Assessment in Education: Principles, Policy, and Practice, 5,* 7–74.

Chen, P.-S. D., Lambert, A. D., & Guidry, K. R. (2010). Engaging online learners: The impact of web-based learning technology on college student engagement. *Computers & Education, 54* (4), 1222–1232.

Clark, D. (2002). Psychological myths in e-learning. *Medical Teacher, 24,* 598–604.
Daradoumis, T., Bassi, R., Xhafa, F., & Caballé, S. (2013). A review on massive e-learning (MOOC) design, delivery and assessment. In F. Xhafa, L. Barolli, D. Nace, S. Venticinque, & A. Bui (Eds.), *2013 Eighth International Conference on P2P, Parallel, Grid, Cloud and Internet Computing (3PGCIC 2013)* (pp. 208–213). Institute of Electrical and Electronics Engineers. ISBN: 978-0-7695-5094-7.
Daza, V., Makriyannis, N., & Rovira Riera, C. (2013). MOOC attack: Closing the gap between pre-university and university mathematics. *Open Learning: The Journal of Open, Distance and e-Learning, 28,* 227–238.
De Witte, K., Haelermans, C., & Rogge, N. (2015). The effectiveness of a computer-assisted math learning program. *Journal of Computer Assisted learning, 31*(4), 314–329.
Galligan, L., & Hobohm, C. (2015). Investigating students' academic numeracy in 1st level university courses. *Mathematics Education Research Journal, 27*(2), 129–145.
Gibbs, A. L. (2014). Experiences teaching an introductory statistics MOOC. In K. Makar, B. de Sousa, & R. Gould (Eds.), *Sustainability in statistics education. Proceedings of the Ninth International Conference on Teaching Statistics (ICOTS9, July, 2014), Flagstaff, Arizona, USA.* Voorburg: The Netherlands: International Statistical Institute.
Hlavac, M. (2015). *Stargazer: Well-formatted regression and summary statistics tables.* R package version 5.2. Online: http://CRAN.R-project.org/package=stargazer.
Jacobson, M., & Archodidou, A. (2000). The design of hypermedia tools for learning: Fostering conceptual change and transfer of complex scientific knowledge. *Journal of the Learning Sciences, 9,* 149–199.
Jungic, V., Kent, D., & Menz, P. (2012). On online assignments in a calculus class. *Journal of University Teaching & Learning Practice, 9*(1), 3.
Karpicke, J. D., & Roediger, H. L. (2008). The critical importance of retrieval for learning. *Science, 319,* 966–968.
Machin, S., McNally, S., & Silva, O. (2007). New technology in schools: Is there a payoff? *The Economic Journal, 117*(522), 1145–1167.
Mayes, J. T., & Fowler, C. J. (1999). Learning technology and usability: A framework for understanding courseware. *Interacting with Computers, 11*(5), 485–497. https://doi.org/10.1016/S0953-5438(98)00065-4.
Nelson Laird, T. F., & Kuh, G. D. (2005). Student experiences with information technology and their relationship to other aspects of student engagement. *Research in Higher Education, 46*(2), 211–233.
Papastergiou, M. (2009). Digital game-based learning in high school computer science education: Impact on educational effectiveness and student motivation. *Computers & Education, 52*(1), 1–12. https://doi.org/10.1016/j.compedu.2008.06.004.
Potocki, A., Ecalle, J., & Magnan, A. (2013). Effects of computer-assisted comprehension training in less skilled comprehenders in second grade: A one-year follow-up study. *Computers and Education, 63,* 131–140. https://doi.org/10.1016/j.compedu.2012.12.011.
Robinson, C. C., & Hullinger, H. (2008). New benchmarks in higher education: Student engagement in online learning. *Journal of Education for Business, 84*(2), 101–108.
Rouse, C. E., & Krueger, A. B. (2004). Putting computerized instruction to the test: A randomized evaluation of a "scientifically based" reading program. *Economics of Education Review, 23*(4), 323–338. https://doi.org/10.1016/j.econedurev.2003.10.005.
Seppälä, M., Caprotti, O., & Xambó, S. (2006). Using web technologies to teach mathematics. In *Proceedings of Society for Information Technology and Teacher Education International Conference, Orlando, FL* (pp. 2679–2684), March 19, 2006.
Sun, P.-C., Tsa, R. J., Finger, G., Chen, Y.-Y., Yeh, D. (2008). What drives a successful e-learning? An empirical investigation of the critical factors influencing learner satisfaction. *Computers & Education, 50*(4), 1183–1202.
Zhang, D., & Nunamaker, J. F. (2003). Powering e-learning in the new millennium: An overview of e-learning and enabling technology. *Information Systems Frontiers, 5*(2), 207–218.

Chapter 6
A Customized Learning Environment and Individual Learning in Mathematical Preparation Courses

Karin Landenfeld, Martin Göbbels, Antonia Hintze and Jonas Priebe

Abstract Large differences in the mathematical knowledge of incoming engineering students as well as different individual students' needs make it necessary not only to offer mathematical preparation courses, but also to conduct these courses in a manner adaptive to the special needs of each student. The online learning environment viaMINT, developed by the Hamburg University of Applied Sciences, accounts for both the student's individual prior knowledge and learning pace as well as the needs due to the course of study. Individualization and customization is realized by several technical and didactical measures, visual elements and a so called short learning track. Students begin by taking an online placement test to assess their knowledge level. Based on the test results, video-based online learning modules, which include various exercises, are recommended. The "Personal Online Desk" gives a customized representation of these recommendations that tracks the student's learning progress. Parallel class lectures explore the topics of the online modules in greater depth to create a blended learning approach.

Keywords e-Learning · Blended learning · Individualized online learning environment · Video-based interactive learning · Online placement test
Preparation courses · Mathematics · STEM subjects · Moodle · STACK

K. Landenfeld (✉) · M. Göbbels · A. Hintze · J. Priebe
Faculty of Engineering and Computer Sciences, Hamburg University
of Applied Sciences, Berliner Tor 7, 20099 Hamburg, Germany
e-mail: Karin.Landenfeld@haw-hamburg.de

J. Priebe
e-mail: Jonas.Priebe@haw-hamburg.de

© Springer International Publishing AG, part of Springer Nature 2018
J. Silverman and V. Hoyos (eds.), *Distance Learning, E-Learning and Blended Learning in Mathematics Education*, ICME-13 Monographs,
https://doi.org/10.1007/978-3-319-90790-1_6

6.1   Introduction

6.1.1   Problems

Many universities offer bridging courses in mathematics for first semester students in STEM fields to close possible knowledge gaps between high school and university. These preliminary courses often have the characteristics of a crash course: during one or two weeks before the semester starts, a large part of high school mathematics is repeated to ensure that students have the background necessary for the mathematical subjects of study courses. In general, during the preliminary course, students work on all topics and in the same way, regardless of their individual mathematical strengths and weaknesses. On the contrary, research shows that students' individual knowledge gaps are very different (Knospe, 2012; Landenfeld, Göbbels, Hintze, & Priebe, 2014). In recent years, many online courses have emerged to enable a more time- and location-independent learning (Biehler et al., 2014; Daberkow & Klein, 2015; Roegner, 2014). However, online courses that are adaptive to the individual needs of the students are not often used.

6.1.2   Measures and Objectives

This chapter introduces an online learning environment, viaMINT, developed at the Hamburg University of Applied Sciences—with its video-based content for mathematics, physics, chemistry and programming—and focuses on the following questions in order to offer an individualized customized learning environment for students:

(1) How is it possible to design the online learning environment and the offered content to take into account the prior knowledge and learning pace of each student? For this, various technical and educational measures are described below.
(2) Our evaluations during the last semesters with the first semester students of the *Faculty of Engineering and Computer Sciences* show clearly that the students welcome the online modules (see Sect. 6.5), but also still want to attend preparation courses at the university. So, how can a meaningful combination of online and classroom teaching be obtained?

6.2   Customized Online Learning Environment

6.2.1   Mass Customization and Individualized Learning in an Adaptable Learning Environment

Within the consumer market, mass customization has become increasingly popular, as it allows customers to define a product that takes into account their personal requirements. The adoption of mass customization into the field of education is introduced in Mulder (2005) as mass individualization of higher education. Individualization of the learning environment and learning process opens new opportunities to conduct an adequate preparation fitting the specific needs of each individual student and her or his course of study.

Individualized Learning (I-Learning) is influenced by multiple elements unique to the learner (for example marks, life situation), the individual learning features provided by the online learning environment and the individual learning behavior of the learner. Zeitler (2016) stated, that it is necessary for future learning to shift from e-Learning to an integrative, individualized, intelligent I-Learning.

Individualized learning is stated to be instructive and teacher-centered, whereas personalized learning is learner-centered, where the student is encouraged to learn in that way that suits his ability and his own preferences (Bray & McClaskey, 2014; Wallach, 2014). viaMINT combines elements of both personalized and individualized learning.

For individualization, viaMINT uses information on the course of study and the results of the online placement test. Based on this information, the learning environment and its learning elements are adapted and individual user-specific learning recommendations are given. In Fig. 6.1, the concept of mass customization for individualized learning is displayed parallel to the concept of individualized learning in viaMINT. Details are described in the next sections. Personalized learning is incorporated by using the different features of the online learning environment viaMINT for individual learning.

6.2.2   The Online Placement Test

viaMINT offers an integrated online placement test in an adapted Moodle[1]-environment. The test demonstrates to the students the mathematical prerequisites for their course of study and gives them the opportunity to preemptively refine their individual skills. Based on the test results, individual learning modules are

---

[1]Moodle is an open source online learning platform: https://moodle.org/.

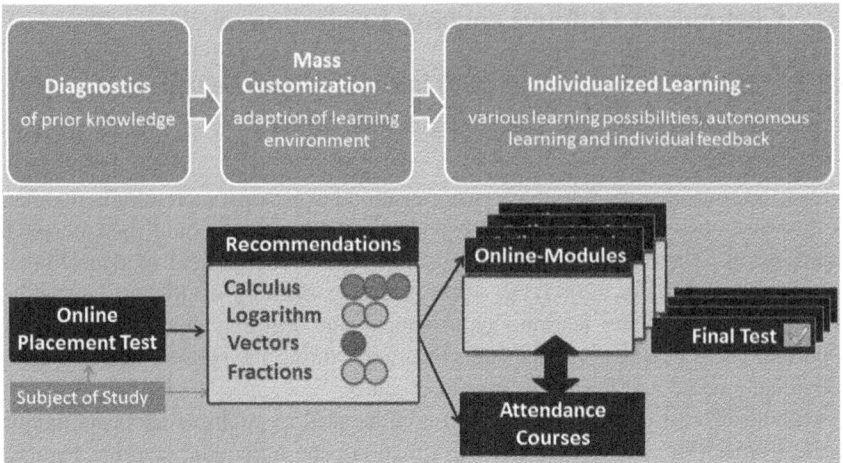

Fig. 6.1 General concept of mass customization for individualized learning in viaMINT

recommended and presented to the students on their *Personal Online Desk* (Landenfeld, Göbbels, & Janzen, 2015).

The online placement test for mathematics consists of ten topics based on school mathematics (see Fig. 6.2) with five questions per topic. The mathematical test topics are: fractions and percentages, equations and inequalities, functions 1 (linear and quadratic functions), functions 2 (polynomial, exponential and logarithmic functions), powers and roots, vectors, logarithms, systems of linear equations, trigonometry and trigonometrical functions and mathematical basics like sets, logic and terms. The topics are chosen from subjects stated in the COSH *Catalogue* (COSH, 2014) to be required by engineering and economics study courses. The topics additionally take into account requirements mentioned by the first semester teachers.

The evaluation algorithm of the online test gives recommendations on a three-step scale: Strongly Recommended, Recommended or Not Necessary (see Table 6.1). After the placement test, Strongly Recommended and Recommended modules appear in the section *Recommended Modules* on the *Personal Online Desk* (see Fig. 6.3).

In this way, students get both summarized and detailed individual feedback on their knowledge along with recommendations for the topics they should focus their efforts on.

The test is designed to last no more than 90 min. Thus, to obtain accurate recommendations, it is necessary to find the right balance between validity and recommendations on the one hand, and test length and the number and type of test

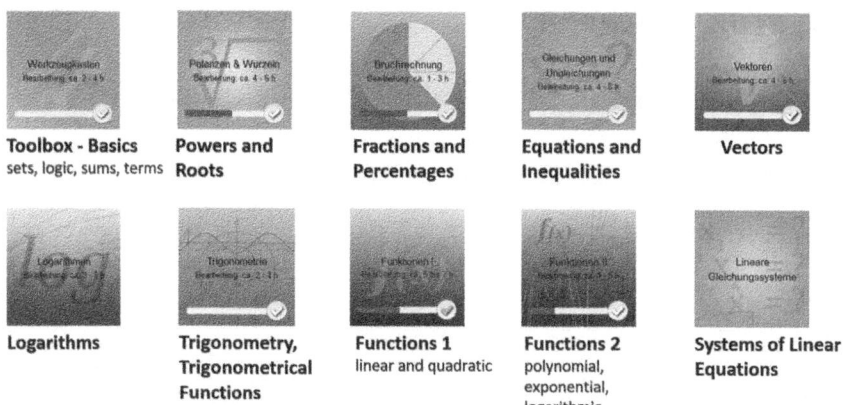

Toolbox - Basics
sets, logic, sums, terms

Powers and
Roots

Fractions and
Percentages

Equations and
Inequalities

Vectors

Logarithms

Trigonometry,
Trigonometrical
Functions

Functions 1
linear and quadratic

Functions 2
polynomial,
exponential,
logarithmic

Systems of Linear
Equations

Fig. 6.2 Mathematical topics in viaMINT used in placement test and learning modules

Table 6.1 Explanation of the three-step recommendation scale

Categories	Correct answers in placement test regarding a particular module (%)	Meaning for the students
Strongly recommended	0–50	Students should work through the entire module from the beginning to the end. They should especially complete all exercises.
Recommended	51–80	Students can focus on single topics of the module. To do so, they have to take a look at the detailed test results to identify the topics of the module they should work on.
Not necessary	81–100	The prior knowledge in this topic is sufficient.

questions on the other hand. Further investigations will deal with this issue of balance. Additional information on the question types is given in Sect. 6.3.

Evaluations of the review courses for the winter semester 2015/16 show that 75.5% of the participating students (total number n = 110) completed the online test, 15.5% partially completed it and 9% not at all. 70.7% of the students who took the test considered it to be helpful or very helpful in realistically estimating their prior knowledge. On average, each module is recommended to more than 50% of the participating students (see Fig. 6.4).

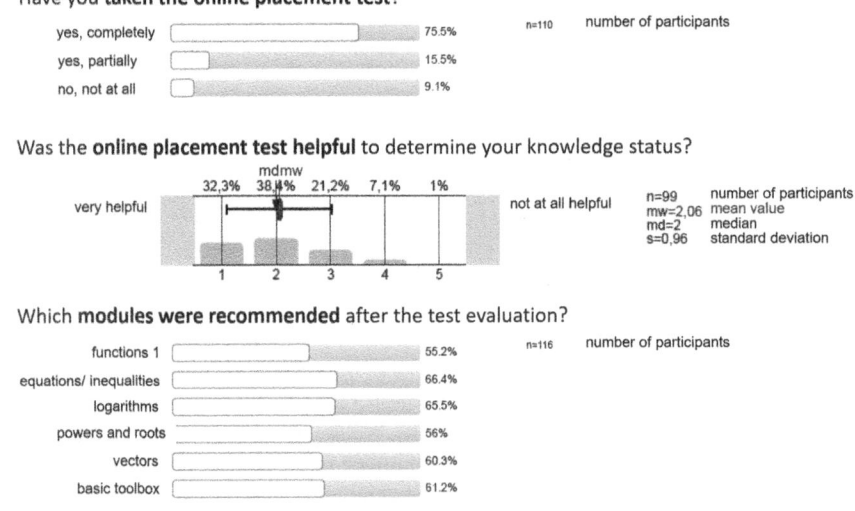

Test Performance

Test-Navigation

1	2	3	4	5	6
7	8	9	10	11	12	13
14	15
21
28
34	35
41	42
48	49

Test Evaluation

Topics Teilgebiet	Success rate Erfolgsrate	Grading Bewertung
Vektoren	30 %	!
Werkzeugkasten	90 %	ok
Potenzen und Wurzeln	85 %	ok
Logarithmen	72 %	ok
Trigonometrie	50 %	!
Funktionen 1	80 %	ok
Gleichungen und Ungleichungen	75 %	ok
Bruch- und Prozentrechnung	92 %	ok
Funktionen 2	35 %	!
Lineare Gleichungssysteme	80 %	ok

Recommendations for Learning

▼ abgeschlossene Module — Completed Modules

▼ belegte Module — Modules in Progress

▼ empfohlene Module — Recommended Modules

Personal Online Desk

Fig. 6.3 Online placement test evaluation and recommendations

Have you taken the online placement test?

yes, completely	75.5%
yes, partially	15.5%
no, not at all	9.1%

n=110   number of participants

Was the online placement test helpful to determine your knowledge status?

mdmw
32,3% 38,4% 21,2% 7,1% 1%

very helpful   not at all helpful

1 2 3 4 5

n=99   number of participants
mw=2,06   mean value
md=2   median
s=0,96   standard deviation

Which modules were recommended after the test evaluation?

functions 1	55.2%
equations/ inequalities	66.4%
logarithms	65.5%
powers and roots	56%
vectors	60.3%
basic toolbox	61.2%

n=116   number of participants

Fig. 6.4 Evaluation results winter semester 2015/2016 concerning the test performance

6.2.3   The Personal Online Desk

The *Personal Online Desk* shown in Fig. 6.5 is divided into different sections: *Recommended Modules, Modules in Progress* and *Completed Modules*. Modules are depicted in sections corresponding with students learning progress, thus organizing the students' learning. An estimated work time is displayed on each module icon. A tooltip shows further information, such as total length of the embedded videos and number of embedded questions. Different visual elements like progress bars, green check marks and badges serve as gamification elements while also giving users a continual feedback on their learning progress.

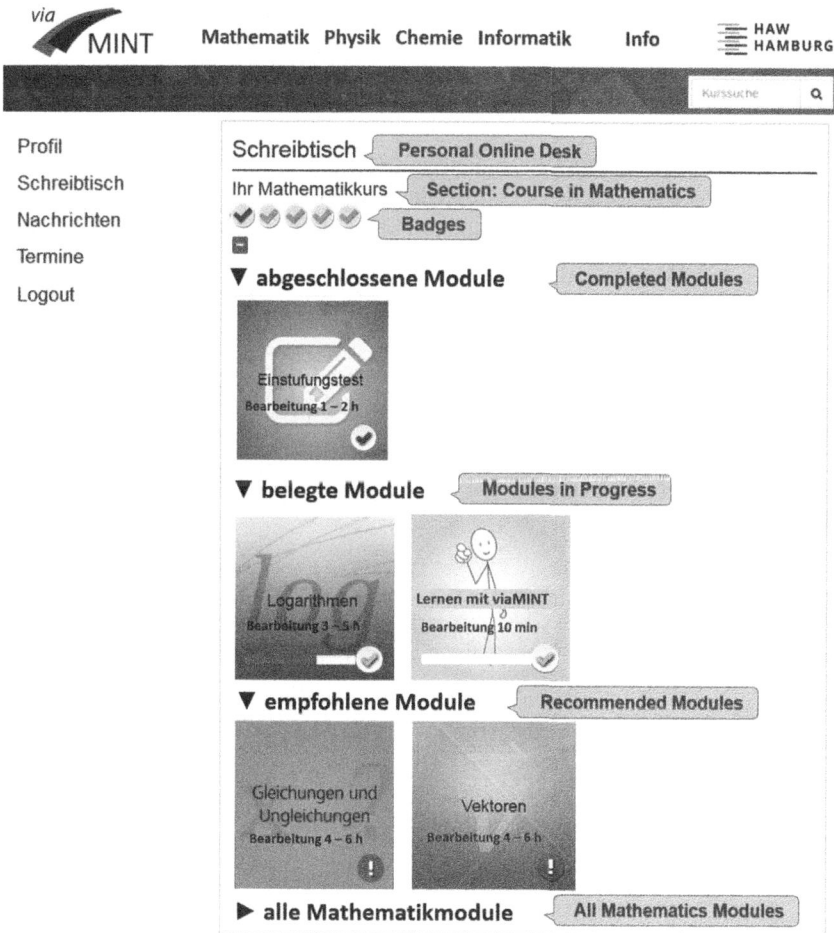

Fig. 6.5 The *Personal Online Desk* with individual settings. viaMINT is in German, so a short English translation is given here

6.2.4 Modules, Learning Sequences and Short Learning Track

Each learning module refers to one of the ten test topics (see Fig. 6.2). The modules (e.g., 'Functions 2') are structured in chapters (e.g., 'Power and Polynomial Functions') and subchapters (e.g., 'Monomial Functions'). Subchapters contain the learning sequences that have an estimated working time of 30–60 min and may also contain additional learning materials like formulary and exercises that approach the topic more in depth (see Fig. 6.6).

The content is mostly presented in videos using a screen capture format, about 6 to 10 short videos concerning on each topic. The sequences of videos are supplemented by interactive applets, questions and online exercises with direct feedback (see Fig. 6.7). Through this approach, a good balance between explanations and hands-on learning is obtained.

To meet the needs students with little prior knowledge, the explanations in the videos are quite extensive. Faster access to the content is given via a *short learning track* (see Fig. 6.8).

In the *short learning track,* students only use the last two elements of each learning sequence. The second-to-last element summarizes the content in a compact video. As shown in Fig. 6.7, this summary video is followed by a summary exercise that students use to check whether they have mastered the content of the learning

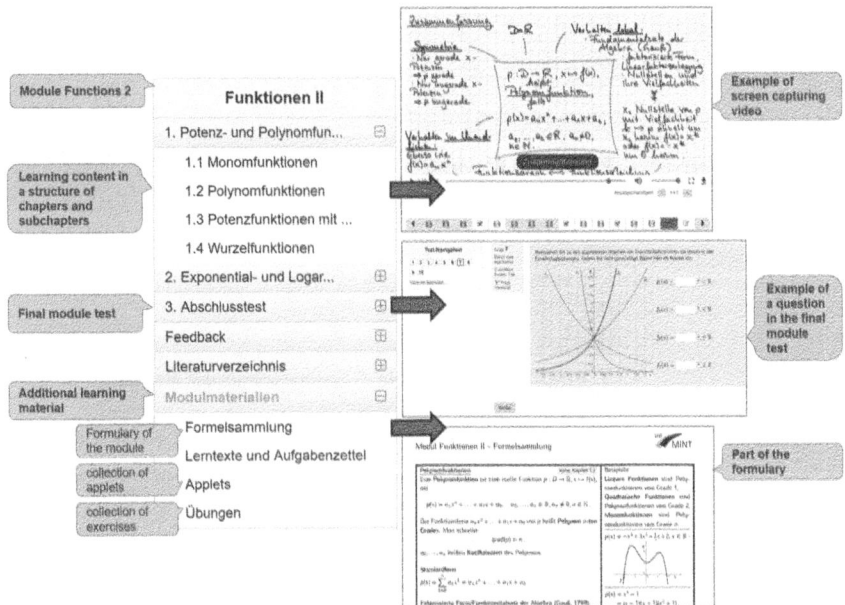

Fig. 6.6 Example of the structure of the module *Functions 2* with brief insight into a learning sequence, the final module test and the formulary

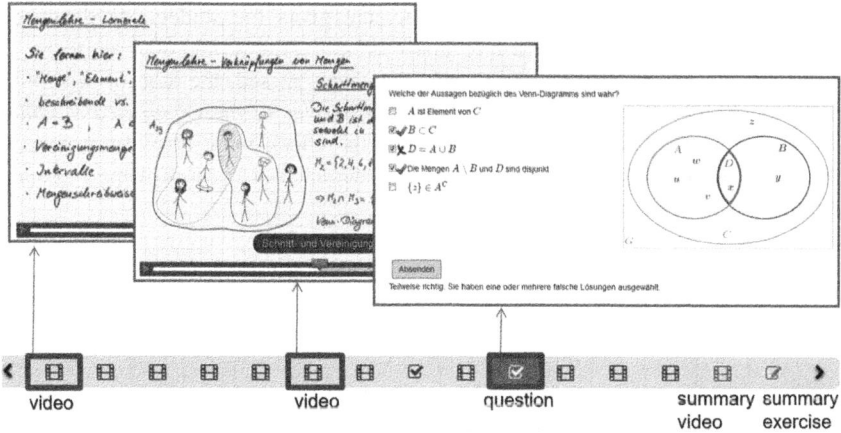

Fig. 6.7 Learning sequence within a sub-chapter, a combination of video explanations and interactive questions

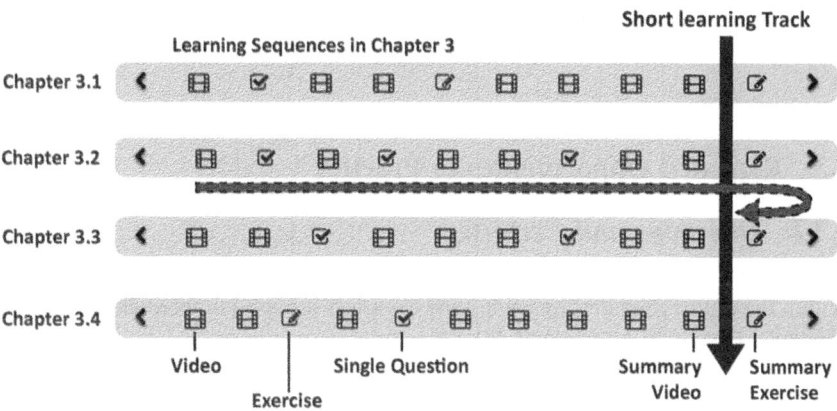

Fig. 6.8 The *Short Learning Track*—individual learning across the learning sequences (red arrow). If desired, more details are given working through the entire learning sequence (grey arrow) (Color figure online)

sequence or not. This way students get a short explanation of the content and corresponding exercises to check their comprehension. In case of incomplete comprehension, students always have the opportunity to work step by step through the whole learning sequence.

In addition to the short learning track, individualization of learning is possible by fast-forwarding or skipping videos. For example, videos with an explanation of an exercise solution may be skipped if the exercise was solved correctly. The last chapter of each module is a final exercise that gives the students feedback on whether they have met the learning goals or not. Answering 80% correctly is equivalent to passing. If the final exercise has been passed, the icon of the corresponding module on the *Personal Online Desk* moves to the *Completed Modules section*. To add a gamification element to it, the check mark turns from grey to green and the user receives a badge which is depicted at the top of the *Personal Online Desk* (see Fig. 6.5).

6.2.5 Individualization Concerning the Field of Study

Each individual course of study may require bridging courses in different fields: mathematics, physics, chemistry and/or programming. The Personal Online Desk is customized depending on the students' individual course of study, displaying only relevant preparation courses (e.g., only mathematics and physics, and no chemistry). The mathematics preparation course itself is also adapted to the needs of the specific study course, i.e. by selecting specific mathematics topics or application examples. The necessary information is taken from the user's profile. With this approach, the amount of preparation courses is adapted to the specific needs of the students. The implementation of this concept is under construction.

6.3 Repeated Opportunities to Practice

6.3.1 Questions and Exercises

Within the online modules, questions and exercises are used in three different ways (Göbbels, Hintze, & Landenfeld, 2014): as brief activating tasks within a learning sequence, as exercise and revision at the end of each learning sequence (so called *summary exercises*, see Fig. 6.7), and as a diagnostic and summative assessment at the end of each module (*final module test*). Different types of questions give the learner feedback and help consolidate the acquired knowledge. The following question types are used in viaMINT: multiple choice with one or more correct answers, drag and drop, numeric and algebraic input, matching and cloze questions. Some question examples are given in Fig. 6.9.

6.3.2 Brief Activating Tasks and Applets

To prevent users from passively viewing the learning videos, two to four of the videos are completed by a short question (see Fig. 6.10). In addition, some of these

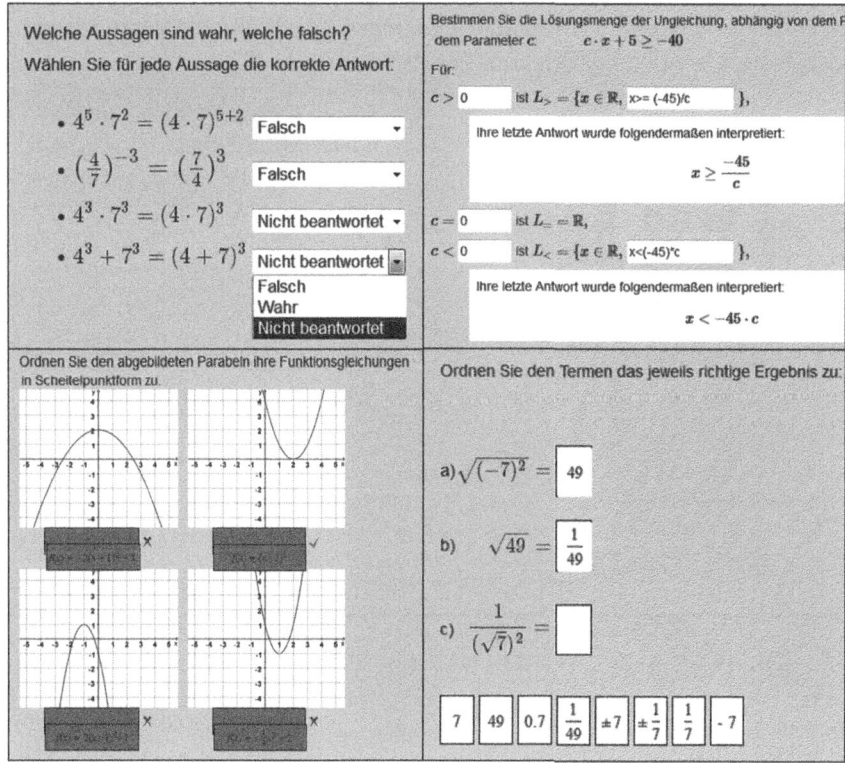

Fig. 6.9 Examples of questions with different question types (drop down menu, drag and drop, algebraic input)

questions involve applets that visualize mathematical relationships and help with answering the questions, as well as with understanding the topic. The input solutions to these questions are evaluated electronically and, if necessary, a detailed explanation is given in the following video. These tasks focus more on understanding the learning content and activating the user and less on training routines and skills.

6.3.3   Exercises and Individual Feedback

At the end of each learning sequence, an exercise set consisting of six to ten tasks is implemented. Here, the goal is to train and consolidate the content discussed in the sequence.

The diversified questions help students reach their learning goals and also serve motivational aspects. Algebraic input questions are realized using STACK/

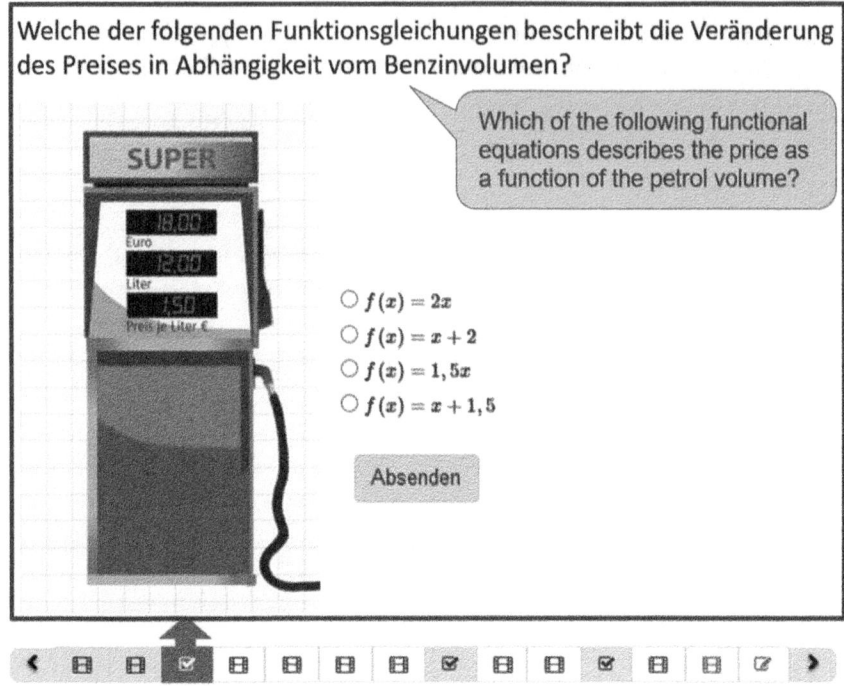

Fig. 6.10 Example for a short question, that follows directly after videos about setting up functional equations

Maxima[2] as a Moodle extension (Sangwin, 2013). Individual feedback is given based on a student's answer. Table 6.2 provides examples of possible responses that students could receive. The program considers typical mistakes and common misconceptions; in this way, student comprehension is maximized (Shute, 2007).

The amount of practice necessary to learn the content differs from student to student. At the moment, about 500 questions are used in the modules and the students are allowed to repeat the exercises without any restriction. Some of the questions use variable values, which are randomized each time the question is accessed and some are randomly drawn from a pool of about 1200 questions. This means students often get a slightly different question every time they open an exercise, allowing them to practice more.

---

[2]STACK—System for Teaching and Assessment using a Computer algebra Kernel http://stack.ed.ac.uk.

Maxima—Computer Algebra System http://maxima.sourceforge.net/.

Table 6.2 Example for individual formative feedback

Problem: Calculate $\frac{1}{6} + \frac{3}{10}$ and express the result in lowest terms. **Expected Answer**: $\frac{7}{15}$		
Answer category	Student answer	Specific feedback
Algebraically equivalent **and** in expected form	$\frac{7}{15}$	**Your answer is correct.**
Algebraically equivalent **not** in expected form	$\frac{28}{60}$ or $\frac{14}{30}$	**Your answer is not reduced to its lowest form.** You can divide by a common factor.
	$\frac{5}{30} + \frac{9}{30}$ or $\frac{10}{60} + \frac{18}{60}$	**Your answer is not reduced to its lowest form.** You haven't added the both correctly expanded fractions.
Not algebraically equivalent **with** particular error or misconception	$\frac{1+3}{6+10}$ or $\frac{4}{16}$ or $\frac{1}{4}$	**Your answer is wrong.** You added numerator and denominator of the two fractions separately.
	$\frac{1 \cdot 3}{6 \cdot 10}$ or $\frac{3}{60}$ or $\frac{1}{20}$	**Your answer is wrong.** You multiplied numerator and denominator of the two fractions.
	$\frac{4}{60}$ or $\frac{4}{30}$	**Your answer is wrong.** You calculated the common denominator correctly, but you haven't expanded the numerators.
Not algebraically equivalent	Something else	**Your answer is wrong.** Calculate the common denominator, add the fractions and then reduce the fraction to its lowest form.

6.3.4   Summative Assessment and Final Test

After the students have gone through a whole module, they can complete the topic by executing the final test for the module. Just like the placement exam, this test has a diagnostic function where the feedback is reduced to whether or not the answers are correct. The questions are generated in a similar way (as for the placement test) and the derived recommendations give the user information about his or her learning status.

6.4   Matching Class Lectures

viaMINT uses a blended learning approach in which the online modules are complemented by corresponding class lectures. Students regard the class lectures as an important part of preparation courses and take advantage of the opportunity to socialize. A combination of online and classroom teaching was conducted for the first time in preparation courses for the winter semester of 2015/16, based on concepts of the *Inverted Classroom Model* (Handke & Sperl, 2012; Loviscach,

2013), commonly also known as *Flipped Classroom Model*. Using this model, it is necessary for the student to work through the online modules before attending the corresponding classroom lectures. The classroom time is spent focusing on practice and consolidation of the online learned content.

Typically, the class lectures start with a multiple-choice quiz using an audience response system (clicker) as a 'warm-up.' The questions refer directly to the content of the associated online module, so the quiz helps the students recall different aspects of the topic and also helps the teacher to identify and correct misconceptions (Göbbels, Hintze, Landenfeld, Priebe, & Stuhlmann, 2016).

The students benefit from the class lectures by further developing their mathematical competencies in modelling, problem posing and solving, and general mathematization of problems. Simultaneously, they are training social and linguistic skills. This is implemented by using open application and modelling tasks and by working with different teaching methods and formats such as group work and experiments.

For example, in the class lecture on *vectors,* students work in groups to solve problems like "How is it possible to sail against the wind?" or "How does a rear reflector work?" (Göbbels et al., 2016). They have to apply their knowledge about vectors learned online, model the situation and after finalization, present their solution to the other groups.

Topics and dates of the class lectures are announced in advance via e-mail. This gives students an idea of whether or not they should participate in the class lecture, depending on their individual learning recommendations and progress.

6.5   Evaluation Results

The learning environment viaMINT and the existing online modules have been developed, deployed, and improved over several semesters in an iterative way. To get detailed information about the students' problems in the field of mathematics, knowledge tests (total number n = 1777) and questionnaires on learning habits and requirements (total number n = 792) were done at the beginning of the project in 2012 and 2013. The results have shown very heterogeneous preliminary knowledge and learning behavior and thus have indicated the need for an adaptive learning environment (Göbbels, Hintze, Landenfeld, Priebe, & Vassilevskaya, 2012).

Since winter semester 2013/2014, when the viaMINT platform and its first online modules were launched, through winter semester 2015/2016, regular evaluations with a total number of n = 878 participants in different student courses have been conducted. The evaluations were carried out during the attendance preparation courses for the Hamburg University of Applied Sciences Faculty of Engineering and Computer Sciences and/or during the mathematics lectures at the start of the semester using an anonymous paper-pencil-feedback questionnaire,

How do you assess the following **technical aspects** of the online platform viaMINT?

▪ **accessibility of the online courses**

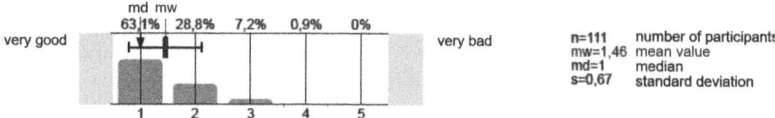

▪ **navigation within the online platform**

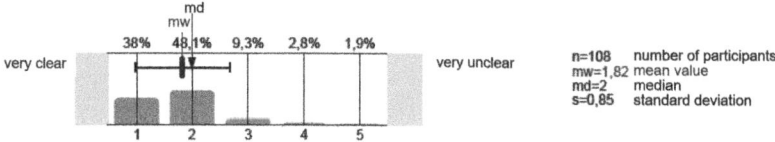

Fig. 6.11 Evaluation results winter semester 2015/2016—technical aspects

which was evaluated with the system EvaSys.[3] The questionnaire contains questions on the technical aspects of the learning environment itself, the didactical approach of the learning modules, and the usability. The results have been evaluated continuously, primarily for improvement of the online learning environment and not for detailed scientific research. The evaluations clearly show that students appreciate the online learning environment with its individual learning opportunities and regard the content as very helpful. Some evaluation results of the survey, especially the technical aspects, the didactical aspects and the learning aspects are shown in Figs. 6.11, 6.12 and 6.13 (Landenfeld, 2016). The evaluation of the technical aspects (see Fig. 6.11) shows clearly a very good accessibility (mean value 1.46 on a scale from 1 to 5) and good usability of the online learning environment with a mean value of 1.82 for the topic navigation.

Figure 6.12 shows the results of the evaluation of didactical aspects related to the tempo of explanation, the comprehensibility and the level of explanation. 47.5% of the participants rate the tempo of explanation as very good. 32% of the participants rate it as a little bit too slow. For these users the short learning track, described in Sect. 6.2.4, has been developed, which offers a faster access to the learning content of the module. The comprehensibility with a mean value of 1.63 on a scale from 1 to 5 makes it obvious that the explanations in the videos, the screen capture format and the exercises are very useful for learning and attests the described approach of viaMINT.

---

[3]EvaSys—webbased software for an automatic evaluation of questionnaires within the scope of quality management www.evasys.de.

How do you assess the following **didactical aspects** during learning with viaMINT?

- **tempo of explanation** within the online modules

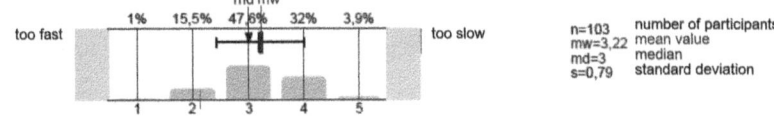

- **comprehensibility**

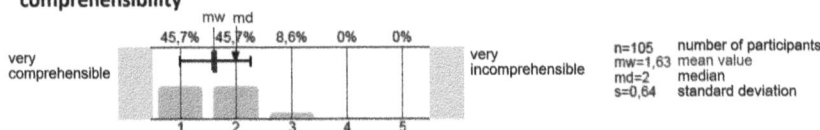

- **level of explanation at begin of module**

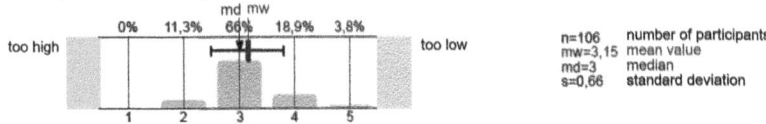

Fig. 6.12 Evaluation results winter semester 2015/2016—didactical aspects

How do you assess **learning with the module logarithm?**

- **quality of learning** with the module logarithm?

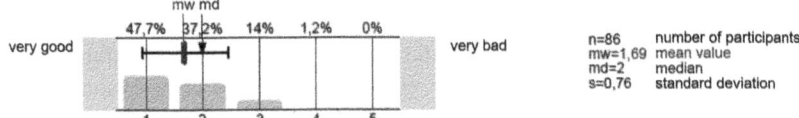

- **time spent for learning** with the module logarithm

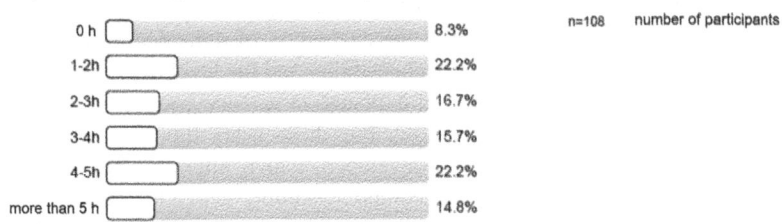

Fig. 6.13 Evaluation results winter semester 2015/2016—learning aspects

The quality of learning and the time spent for learning using one specific learning module are addressed in Fig. 6.13. The quality of learning is rated homogenously with a mean value of 1.69 on a scale from 1 to 5. On the contrary the time spent for learning is rated very inhomogeneously and differs from 0 to more

How do you assess **learning with the module logarithm?**

* How do you rate **your prior knowledge of logarithms - before** working with the module?

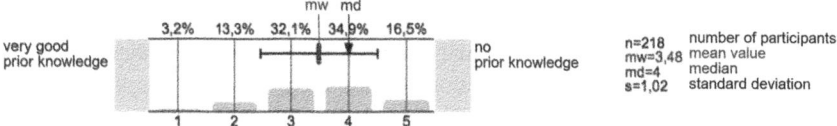

* How do you rate **your knowledge of logarithms - after** working with the module?

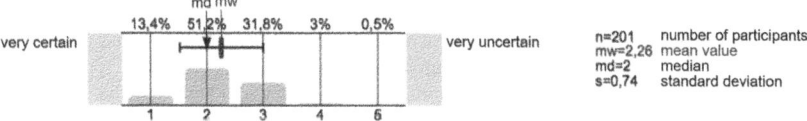

Fig. 6.14 Evaluation results winter semester 2013/2014—self rating of knowledge improvement

than 5 hours distributed nearly uniformly. This demonstrates the very heterogeneous prior knowledge and individual learning behavior.

The impact on the students' performance after learning with viaMINT can be demonstrated in several ways. First, if the learner has passed all final module tests of the recommended modules and received the module badges, he or she has very good knowledge of the module content. Second, the self-rating of the learner after working through the module is shown in Fig. 6.14. It displays, that the learner is confident that he or she has more knowledge and understanding of the content of the module. Since the evaluations were conducted anonymously, it is very difficult to measure student performance within the mathematics lecture in the first semester in relation to the results of the online learning environment. This should be part of a scientific work in future.

6.6   Summary and Outlook

Within this paper the need for individualization in mathematical preparation courses was identified and viaMINT was suggested as a potential customized learning environment with different specific features. Two main elements of individualization are the online placement test with recommendations, which are displayed on the Personal Online Desk, and the short learning track, which allows individual learning in a clear way. Three different ways of assessment using different question types have been pointed out: brief activating tasks, summary exercises and diagnostic and summative assessment. Various types of questions give the learner individual feedback and help to consolidate the acquired knowledge. The blended learning approach of the preparation courses combines the advantages of online learning and in class lectures and deepens the mathematical comprehension.

We are in the process of extending the learning environment with further modules in mathematics and physics, new courses in chemistry and programming will also be developed. The comprehensive evaluation will be continued. Furthermore, the blended learning concepts will be expanded in the next semesters, adapting even further to the individual needs of students and evaluated in detail. The implementation of individualization of the Personal Online Desk concerning the field of study will be realized. Further investigations will also be carried out to find the right balance between validity and recommendations of the online placement test on the one hand and test length and number of test question on the other.

Acknowledgements This work is part of the research developed under the Project "Lehre lotsen" funded by the BMBF (Bundesministerium für Bildung und Forschung), project number 01PL11046. Responsibility for the contents of this publication is assumed by the authors.

References

Biehler, R., Bruder, R., Hochmuth, R., Koepf, W., Bausch, I., Fischer, P. R., et al. (2014). VEMINT – Interaktives Lernmaterial für mathematische Vor- und Brückenkurse. In I. Bausch, R. Biehler, R. Bruder, R. Hochmuth, W. Koepf, S. Schreiber, & T. Wassong (Eds.), *Mathematische Vor- und Brückenkurse – Konzepte, Probleme und Perspektiven* (pp. 261–276). Wiesbaden: Springer Spektrum.

Bray, B., & McClaskey, K. (2014). *Personalization v differentiation v individualization (PDI) Chart (Version 3)*. Online report. http://www.personalizelearning.com/2013/03/new-personalization-vs-differentiation.html. Accessed July 15, 2017.

COSH. (2014). *Cooperation Schule: Hochschule - Mindestanforderungskatalog Mathematik (Version 2.0) der Hochschulen Baden-Württembergs für ein Studium von WiMINT-Fächern.* Ergebnis einer Tagung vom 05.07.2012 und einer Tagung vom 24.-26.02.2014; Stand 23. Juli 2014. Online: http://www.mathematik-schule-hochschule.de/images/Aktuelles/pdf/MAKatalog_2_0.pdf. Accessed July 15, 2017.

Daberkow, M., & Klein, O. (2015). Geht doch – Mathematik online für Erstsemester. In *Tagungsband zum 2. HDMINT Symposium 2015, Nürnberg* (p. 242).

Göbbels, M., Hintze, A., & Landenfeld, K. (2014). Verschiedene Formen intelligenten Übens in mathematischen Online-Kursen. In *Proceedings at: Hanse-Kolloquium zur Hochschuldidaktik der Mathematik 2014, 7./8.11. 2014, Münster.*

Göbbels, M., Hintze, A., Landenfeld, K., Priebe, J., Stuhlmann, A. S. (2016). Blended learning for mathematical preparation courses—Video based learning and matching in-class lectures. In *European Society for Engineering Education, Proceedings of the 18th SEFI Mathematics Working Group Seminar, Gothenburg, Sweden* (pp. 93–98), June 27–29, 2016.

Göbbels, M., Hintze, A., Landenfeld, K., Priebe, J., & Vassilevskaya, L. (2012). Blended Learning für Mathematik-Vorkurse - Eine Bestandsaufnahme der Vorkenntnisse. In J. Vorloeper (Ed.), *Proceedings of 10. Workshop Mathematik in ingenieurwissenschaftlichen Studiengängen 2012, Mülheim/Ruhr* (pp. 25–34).

Handke, J., & Sperl, A. (Ed.). (2012). Das inverted classroom model. In *Begleitband zur ersten deutschen ICM Konferenz.* Münster: Oldenbourg.

Knospe, H. (2012). Zehn Jahre Eingangstest Mathematik an Fachhochschulen in Nordrhein-Westfalen. In J. Vorloeper (Ed.), *Proceedings of 10. Workshop Mathematik in ingenieurwissenschaftlichen Studiengängen 2012, Mülheim/Ruhr* (pp. 19–24).

Landenfeld, K. (2016). viaMINT: Videobasierte interaktive Vorkurse – Eine Online-Lernumgebung für den Studieneinstieg im Blended Learning-Format an der Fakultät

Technik und Informatik. Collective Report, In *Lehre lotsen 2011–2016 – Erste Förderphase Dialogorientierte Qualitätsentwicklung für Studium und Lehre an der Hochschule für Angewandte Wissenschaften Hamburg* (pp. 73–88). HAW Hamburg. ISBN 978-3-00-054734-8.

Landenfeld, K., Göbbels, M., Hintze, A., & Priebe, J. (2014). viaMINT – Aufbau einer Online-Lernumgebung für videobasierte interaktive MINT-Vorkurse. *Zeitschrift für Hochschulentwicklung (ZFHE), 9*(5). Online: www.zfhe.at/index.php/zfhe/article/view/783/642.

Landenfeld, K., Göbbels, M., & Janzen, S. (2015). Der Persönliche Online-Schreibtisch in der Vorkurs-Lernumgebung viaMINT. In *Poster publication and Proceedings at GMW & DeLFI 2015 „Digitale Medien und Interdisziplinarität: Herausforderungen, Erfahrungen und Perspektiven".* Online: http://www.interdis2015.de/poster.html. Accessed July 15, 2017.

Loviscach, J. (2013). MOOCs und Blended Learning – Breiterer Zugang oder Industrialisierung der Bildung? In R. Schulmeister (Ed.), *MOOCs—Massive open online courses—Offene Bildung oder Geschäftsmodell* (pp. 239–255). Berlin: Waxmann.

Mulder, F., (2005). Mass-individualization of higher education facilitated by the use of ICT. In J. M. Haake, U. Lucke, D. Tavangarian (eds.), *DeLFI 2005: 3. Deutsche e-Learning Fachtagung Informatik* (p. 17). Bonn: Gesellschaft für Informatik. Online: http://dl.mensch-und-computer.de/handle/123456789/1748. Accessed July 15, 2017.

Roegner, K. (2014). Exploratives Lernen an der Schnittstelle Schule/Hochschule. In I. Bausch, R. Biehler, R. Bruder, R. Hochmuth, W. Koepf, S. Schreiber, & T. Wassong (Eds.), *Mathematische Vor- und Brückenkurse – Konzepte, Probleme und Perspektiven* (pp. 181–196). Wiesbaden: Springer Spektrum.

Sangwin, C. (2013). *Computer aided assessment of mathematics.* Oxford: Oxford University Press. ISBN: 9780199660353.

Shute, V. J. (2007). Focus on Formative Feedback. *ETS Research Reports Series, 2007*(1), i-47. https://doi.org/10.1002/j.2333-8504.2007.tb02053.x.

Wallach, S. (2014). *Personalized learning vs. individualized learning.* Internet report on Edmentum Online. Online: http://blog.edmentum.com/personalized-learning-vs-individualized-learning. Accessed July 15, 2017.

Zeitler, W. (2016). Humboldt digital: E-learning oder I-learning? In *DNH – Die Neue Hochschule, Heft 2/2016* (pp. 46–47). Online: http://hlb.de/fileadmin/hlb-global/downloads/dnh/full/2016/DNH_2016-2.pdf. Accessed July 15, 2017.

Part III
Innovations on E-Math
Learning and Teaching

Chapter 7
Scripting Collaboration for Competence-Based Mathematics Learning: A Case Study on Argumentation

Giovannina Albano and Umberto Dello Iacono

Abstract This work concerns the use of scripting collaboration in order to implement an innovative approach to competence-based mathematics learning. In this chapter, we show how Digital Interactive Storytelling in Mathematics (DIST-M) can be used to engage students in shared argumentative experiences and how the technology supports the students' reprocessing and appropriation within their knowing. The design is based on a network of theories and students are engaged in activities within a storytelling experience. The activities use both experiential and discursive approaches to mathematics learning, integrating individual and social tasks, defined by external scripts. We merge free tools and define new applications, which allow to integrate and manipulate dynamic graphs as well as to construct open sentences starting from available blocks of words. We also discuss the outcomes of a case study.

Keywords Mathematics education · Collaboration script · Digital storytelling Argumentation · Linguistic cohesion

7.1 Introduction

This paper concerns the definition of a design methodology, Digital Interactive Storytelling in Mathematics (DIST-M), for competence-based mathematics learning in e-learning environment. It is based on the assumption that such an environment can be arranged in a way that a good exploitation of platform tools and a well-structured collaboration among peers can act as an "expert" and scaffold

G. Albano (✉) · U. Dello Iacono
c/o DIEM—University of Salerno, Via Giovanni Paolo II, 132,
84084 Fisciano, SA, Italy
e-mail: galbano@unisa.it

U. Dello Iacono
e-mail: udelloiacono@unisa.it

© Springer International Publishing AG, part of Springer Nature 2018 115
J. Silverman and V. Hoyos (eds.), *Distance Learning, E-Learning and Blended Learning in Mathematics Education*, ICME-13 Monographs,
https://doi.org/10.1007/978-3-319-90790-1_7

students in achieving their learning goal (Albano, Dello Iacono, & Fiorentino, 2016; Albano, Dello Iacono, & Mariotti, 2017). The underpinning theoretical approach is framed in the socio-constructivist view of learning, where students construct their own knowledge and are actively engaged in social interactions (Vygotsky, 1978). The DIST-M consists of collaboration scripts, aimed at regulating and structuring roles and interaction in a collaborative setting (King, 2007). As suggested by the name, the described methodology is implemented in a story-telling framework, where the students are characters in a story and they interact facing problems, whose solution is needed to advance. The design on one hand can motivate learners and, on another hand, can have benefits of the integration between narrative and logical-scientific thought (Zan, 2011).

In this chapter, we apply the DIST-M in a case study concerning the argumentative competence in mathematics and analyze whether it can promote the production of written arguments according to a register shared in the mathematical scientific community. In fact, as shown by PISA results, a critical challenge for 15 years old students is expressing arguments and conclusions in written form (Turner & Adams, 2012). In the frame of discursive approach to mathematics learning, seen as initiation to a particular mathematical discourse (Sfard, 2001), Ferrari (2004) shows that mathematical language and written literate registers of ordinary language share many features. Thus, he concludes that being familiar with written communications is a prerequisite to promote advanced mathematical thinking. To this aim, there is a need of a shift from the request of just solving a problem to the request of verbal and or written explanations.

In the following, first, we share the theoretical framework, then we describe the design of DIST-M. Finally, in the case study we discuss its outcomes from a qualitative point of view, taking into account the collaborative and argumentative features of the design. The analysis will highlight the arguments produced by the students and assumes a linguistic perspective, focusing on the organization of the verbal texts, as cohesive texts, which means words and sentences perceived as a whole entity.

7.2   Theoretical Framework

This work approaches mathematics learning, taking into account a network of theories: discursive approach (Sfard, 2001), computer supported collaboration script (King, 2007), especially with respect to argumentative competence (Weinberger, Stegmann, Fischer, & Mandl, 2007).

Sfard's (2001) notes that, communication in mathematics cannot be considered simply an aid to thinking, but, primarily because mathematics learning can be defined as initiation to a particular (mathematical) discourse, it should be thought of as thinking. When students construct arguments, they elaborate and explain to themselves the concepts that they are justifying (Baker, 2003). Such explanations to themselves help the students to integrate new information within cognitive structures previously existing (Chi, Bassok, Lewis, Reimann, & Glaser, 1989). When students share their explanations in a work group, they are expected to produce arguments, in form of communicable texts (Boero, 1999), that are socially acceptable, and, in particular, "mathematically acceptable" (Mariotti, 2006). In the frame of the discursive approach to mathematics learning (Sfard, 2001), the quality of such texts is strictly linked to the quality of thinking, and thus of argumentation, so that the use of literate registers should be considered an indicator of quality, and at the same time an educational objective (Ferrari, 2004).

In spite of the benefits of collaborative work, it is well known that collaboration is not spontaneous and successful without being well structured (Laurillard, 2013). To this aim, educational scripts can be externally imposed, regulating roles and actions that the students are expected to assume and to carry out in order learning occurs successfully in collaborative learning (King, 2007). Their use has been implemented in computer-based environments (Weinberger, Kollar, Dimitriadis, Mäkitalo-Siegl, & Fischer, 2009), where external scripts can establish the roles of the participants and the sequence of the tasks to be performed (King, 2007), regulating the interaction and collaboration in order to foster suitable cognitive processes. Two levels can be distinguished in a script: one macro and one other micro. The former concerns how to group and to assign the roles, whilst the latter defines a suitable sequence of events and of specific tasks to be carried out for the effectiveness of the cooperation. The aim of the educational scripts is that they will be interiorized along the time through the social practice (Vygotsky, 1978), in order to bring the students to be more autonomous learners.

Literature also shows the failure of spontaneous collaboration in order to produce arguments (Kuhn, Shaw, & Felton, 1997). Andriessen, Baker, and Suthers (2003) notes that computer based collaboration scripts can help, since they allow the students to communicate among them by means of text-based interfaces and to write and read texts and messages. This modality let them to be faster than in face-to-face setting in order to read and revise their own speech and the peers' ones (Pea, 1994). Moreover, it also encourages weak students to participate in debates, supporting the recovery of previous gaps.

Weinberger et al. (2007) distinguishes three types of components into a computer based collaboration script on argumentation: epistemic, argumentative, social. The epistemic components aim to structure the collaborative activities, focusing on the discussion's content and on the steps needed to carry out the task. They can support students in finding suitable solving strategies. The argumentative components aim to support the construction of arguments admissible with respect to a fixed formal system. The social components define the interactions among the students in order to promote the construction of knowledge. They take care of the engagement of the students in collaborative activities that cannot occur spontaneously, as for instance encouraging students to reply critically to their peers' contributions.

7.3   The DIST-M Methodology

The methodology defined in Distance Interactive Storytelling (DIST) is competence oriented, framed in an e-learning environment. DIST makes use of digital story-telling, which becomes interactive due to the use of applications allowing the student to manipulate objects (graphics, multimedia, etc.) and due to feedbacks given by the e-learning platform. When a DIST concerns a mathematical compe-tence, it will be called DIST-M, short for Digital Interactive Storytelling in Mathematics.

The DIST is organized as collection of Frames. Each Frame is sequence of scripts and each script consists in one or more tasks, where a task is a learning activity. The first Frame, called Introduction, aims to let the student become familiar with the digital environment, the storytelling and the content pre-requisites. All the other Frames, labelled as "Frame of Level," aim to mediate various levels of the specific competence at stake. Thus, starting from a Frame of Level 1, focused on a basic level of competency, Frames of greater level mediate the same competence at higher levels.

The tasks can be individual, collaborative or mixed. In the individual tasks, the student is expected to work and to delivery her products individually, not com-municating with peers. In the second, collaboration among peers is guided by the task design and it is realized by means of constraints in the use of tools such as chat, forum, wiki, etc. In the mixed tasks, the student can take advantage of communi-cation with peers (usually by chat) but she is required to delivery her work indi-vidually. The underpinning idea is that scripts should be designed as sequence of collaborative and individual tasks so that learning is first socialized and then interiorized, consistent with a Vygotskian view (Vygotsky, 1978).

7.4   The Design of a DIST-M Concerning Argumentation

In the following, we describe a DIST-M on argumentative competency. The DIST-M aims at providing the students with a methodology of construction and communication of arguments in mathematics. In the literature, some authors studied the effectiveness of collaborative scripts in fostering mathematical argumentation skills (Kollar, Ufer, Reichersdorfer, Vogel, Fischer, & Reiss, 2014; Vogel, Kollar, Ufer, Reichersdorfer, Reiss, & Fischer, 2015) focusing on the construction of a conjecture and on its proving process by means of explanation, arguments, coun-terarguments and synthesis. In our case, we focus on the elaboration of a com-municable text (Boero, 1999) as an answer to a question, explaining the correctness of the answer based on mathematical arguments and expressed in a literate register (Ferrari, 2004) and according to shared socio-mathematical norms which makes the statement acceptable in the reference scientific community (Mariotti, 2006). The design of the scripts within the DIST-M is based on the idea of transferring the mediation role of the teacher to the peers and to the device. Thus, it foresees various types of interactions, both with the device and the peers, and it enables students to produce personal arguments, to compare with the peers' ones and to elaborate their own final argumentative text in a literate register. In this view, the epistemic components of the script do not explicitly give solving strategies, but they aim to recover previous knowledge and skills gaps (see Sect. 7.4.2) and to activate processes of individual and collaborative reflection, whose side-effects should be self-regulation and self-correction. The argumentative components of the DIST-M aim to let the student explain her reasoning to her peers and then to convert it in a literate register. The social components of the DIST-M have been designed in order to promote the interaction among peers, for instance in order to agree upon a common answer and statement, to ask and give help to peers in troubles, and so on.

The actual implementation of the DIST-M includes the Frame Introduction, consisting of one script, and the Frame of Level 1, consisting of three scripts, named Chapter 1, Chapter 2 and Chapter 3. It has been implemented in Moodle, a common e-learning platform (https://moodle.org). Among its tools, we have used: Chat (for informal and speed communication); Answer and Question Forum (as students cannot view other students' posts before posting their own comment); Wiki (as a LogBook, where to write information or add images useful for the continuation of the activities); Lesson Module (to allow personalized learning path according to the student's needs along the DIST-M); Task Module (to share the students' products within a group). Further, we have integrated Moodle pages with comic strips, realized by Tondoo (www.toondoo.com), to implement the story-telling, and with new interactive manipulative objects, realized by GeoGebra (www.geogebra.org) as shown in the next section.

7.4.1 The Technological Innovations

Three new resources, consisting of interactive applications created by GeoGebra, have been defined: Tutorial, Interactive Graphical Question and Interactive Semi-Open Question. Once created, they have been loaded to the Community for the users of GeoGebra (https://www.geogebra.org/materials) and they have been made visible by the URL address and the HTML code has been incorporated in the short answer question page (a single word or short phrase answer should be provided) of the Moodle Lesson. In the following chapters, we will detail each resource.

The Tutorial enables the student to inteact with a designed object (graphs, text boxes, tables, etc.) in order to find a configuration of the object as answer to a given question. The application, at the end of the student's manipulation, gives back a code, according to the correctness or not of the given answer. The student is expected to insert, in a suitable text box of the short answer of the Lesson page, such code, which steers the student to a subsequent personalized path.

The Interactive Graphical Question (IGQ) differs from Tutorial, since the code given back does not correspond only to the two options (correct or incorrect answer). Given that in many cases, various decisions and configurations of the graphical objects can be correct, the IGQ allows for a variety of parameters, so each configuration can be correct (all parameters are admissible), semi-correct (some parameters are admissible), wrong (no parameter is admissible). Thus, the application has been designed in order to be able to generate dynamically a code that reveals the manipulation of the student with respect the two previous features. Then the code allows a very fine subsequent personalization.

The Interactive Semi-Open Question (ISQ) allows to construct the answer to a given question by assembling some available words-blocks by means of dragging. The expected answer should be constituted as a main sentence linked to a secondary one, which concerns the arguments to support what stated in the main sentence. The correctness of the answer depends on the two sentences, thus we can have correct, semi-correct and wrong answers. The code given back is generated dynamically taking into account which words-blocks have been used and how they have been assembled, allowing to know exactly which sentence the student created and thus to foresee fine personalized paths. From the technological point of view, the ISQ allows the environronment to overcome, at least partially, the problem of automatic assessment of open-ended questions. Actually, it can be very near to a real open-ended question if the words-blocks are suitably chosen to allow the student to construct sentences alike in language and thought to the ones she actually uses in a similar situation. From the educational point of view, the careful selection of the words-blocks to be made available can foster the argumentative competence. In fact, in our case, they can make evident the general structure of an argumentation in a literate register, highlighting the causal conjunction between the main sentence and the subordinate one, independently on the order of the two sentences (Albano, Dello Iacono, & Mariotti, 2017).

7.4.2 The Frame Introduction

The Frame Introduction aims to steer the student within the storytelling and the digital environment. It consists of a single script that introduces the student in the story "Discovery Programme". As an example, on Discovery Programme noted "Life on the planet Terra is at risk: a huge impact with a meteorite is foreseen in 2150, which can cause the extinction of life on the planet. All the scientists in the world are working hard in order to find a solution as soon as possible. A new planet in the solar system has been discovered and the NASA has launched the space probe COLOMBO to collect data from the new planet, which need to be analyzed in order to test if life is possible on it." In this case, the student plays the role of a scientist of the NASA, as member of an équipe (team) supervised by Professor Garcia (guide and voice of the storytelling) (Fig. 7.1).

Nella sede di Houston, Texas, il Prof. Andrew Garcia guida una equipe di 4 scienziati, i matematici americani Peter Martin e Anna Clark, la biologa italiana, Daniela Rocchi e il geologo italiano, Giorgio Pisani. L'obiettivo dell'equipe è quello di analizzare e rappresentare in tempo reale i dati provenienti da COLOMBO.

Fig. 7.1 The équipe in discovery programme

Along the storytelling, the student is going to face questions concerning statistics and statistical graphics, needed for the équipe work. Thus, the mathematical content at stake concerns representation and management of graphics of descriptive statistics, which constituted the contents of the Tutorials (which are not the focus in this paper).

7.4.3   The Frame of Level 1

In the following, we go into details of the script Chapter 1 within the Frame of Level 1. The design, shown in Fig. 7.2 is based on both experiential and discursive approach. So, on one hand, the student can manipulate interactive objects in order to formulate and test hypotheses and, on the other hand, she is expected to debate with herself and with the peers.

At beginning (task 1) each student is expected to choose a role to play in her group, by negotiating it with the colleagues. The roles foreseen are the following:

- the Captain, the leader of the group who takes care of engaging all the mates in the discussions and in the decision processes (social literacy);
- the Scientific Official, who is in charge of collecting and summarizing all the mates' answers concerning mathematical questions to be solved during the mission (mathematical literacy);

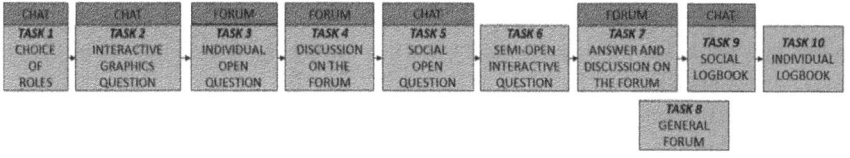

Fig. 7.2 Design of script Chapter 1

- the Technological Official, who supports the mates who are in troubles in using the platform (digital literacy);
- the Communication Official, who reports and summarize the conversations of the mates when a shared communication/answer is required (communicational literacy).

Then, the group is required to work with an ISQ, where suitable manipulation of a graphical object gives the answer to a posed question (task 2 part 1). According to the code, corresponding to the final object's configuration chosen by the student (see Sect. 7.5.2), a personalized reflective question is delivered (task 2 part 2). It aims to steer the student towards self-regulation processes by means of being aware of what done and why: further possible correct configurations of the object, if she was successful; reasons of her choice, in case of semi-correct configuration; and on what has brought her to generate a wrong configuration, in case of failure.

Task 3 requires the student to answer to an individual open-ended question, aimed to shift from the previous experience to a general case. It should bring out the elaboration of arguments to justify the given answer. Each student has to post her answer in a Question and Answer Forum, in order to avoid the influence of peers' arguments and to force the participation.

When all the students posted their answers, in order to elaborate a shared answer (task 4), a discussion is started in the same Forum. The use of the forum guarantees that everyone completes the previous task, otherwise they cannot access to the peers' answers, and differing from the chat where communication is immediate and not so formal, in the forum there is an implicit request of a shift towards a more literate register. Once agreed upon the answer, the students deliver it by the Moodle Task module, in collaborative setting so that each member takes her responsibility to deliver the shared answer (task 5).

Later (task 6), the student individually converts it into a more literate text, assembling suitable words-blocks (see ISQ, Sect. 7.4.1). The words-blocks have been constructed in order to highlight the causal structure of the sentences, that is the causal conjunctions (i.e. *since*, *because*, etc.) constitute single blocks, allowing to link two sentences (main and conditional ones) constructed by more other blocks. The ISQ recognizes the correctness of the construction independently on the order of the sentences.

The task 7 requires the student to post the answer constructed in the Question and Answer Forum together with explanation of her reasoning and to discuss all the answers in the thread. If she was successful in the previous task, then she is acknowledged of this (she has the title of Champion) and she is asked to help her classmates. A further general Forum is activated where all the Champions (eventually, also a teacher who can be essential when there is no champion) are available for anybody at risk (task 8).

At the end of the activity, the students are required to edit a Social LogBook (task 9) and an Individual LogBook (task 10). The first one is composed using a Moodle collaborative wiki, aiming to collect and store all the cognitive information useful for the mission. The second one is referred to a metacognitive reflection of the student on the activity, on the difficulties encountered and how she overcome them.

The various tasks of the script refer to components that can be social, epistemic and argumentative. All the tasks that envisage debate, discussion, comparison among the students can be ascribed to social components (tasks 1, 4, 5, 7, 8, 9). As the aim of the script is to foster argumentation, most of the tasks encourage the students to reflect in order to explain, to clarify, to produce arguments, so they can be considered argumentative components. Finally, as the arguments the students should produce concern problem solving activities, solving strategies, recovery of errors, many tasks can be seen as epistemic components (tasks 2, 3).

7.5 The Case Study

7.5.1 The Mathematical Problem

The mathematical problem of the case study concerns the invariance of the angle of a circular sector with respect to the length of the radius of the circle. The tasks of the script aim to put the student's attention on the relationship between the size of the angle and the radius of the circle. In task 2, a circle with unit radius is presented, where a circular sector corresponding to a 72° angle represents the percentage of red stone on the new planet (20%) and the student must enlarge the chart manipulating the radius and/or the size of the angle (by means of an IGQ), unchanging the percentage.

After manipulating the circle graph, a scientist of another group appears asking the student a question: "Can you explain how you chose the angle?", in case of semi-incorrect or incorrect configuration, or "I would like to take as radius X and to leave the angle at 72°. Does anything change?", in the case of a correct configuration with a radius different than X. The student faces the same question several times in order to generalize the experience and the results achieved: "How does the size of the angle of colored sector changing the radius? Explain your answer", first individually (task 3), then collaboratively (task 5) and finally by means the ISQ (task 6).

7.5.2 Methodology

The Frame Introduction and Chapter 1 of the DIST-M have been tested in a pilot involving eleven 10th Grade students from Grammar High School "Virgilio" in Southern Italy. The students were randomly split into 4 groups, 3 of which consisting in 3 members and 1 consisting in 2 ones. In groups of 3 members, one of the students played 2 roles (task 1), whilst in the group of 2 members, each student played 2 roles. Each student worked on her PC, for twenty hours, logged into the platform by username (such as S1, S2 etc.) and password, and communicated within a group only by means of the tools of the platform.

7.5.3 Analysis Tools

Our DIST-M aims to support the production of written verbal arguments by the students. Some theoretical models of analysis of the arguments do not refer to language, but we assume that a written argument is, firstly, a written text and the production of a correct text is closely intertwined with an acceptable explanation. This is why we chose to use a linguistic approach to analyze the data, using tools such as textual cohesion that consider the text a single entity rather than a collection of words and disorganized sentences (Halliday & Hasan, 1976). Cohesion is different from coherence, although they are very interrelated. The coherence identifies the connection among the various sentences and, thus, allows to give them continuity of sense, that depends on who writes and who interprets the text, starting from her encyclopedic knowledge. Therefore, coherence is closely linked to the interlocutors rather than to the language itself. Instead, the cohesion concerns the grammatical way in which the sentences are related to each other. Then it refers to the linguistic tools needed to achieve coherence and helps to highlight it, although a text can be perceived as coherent without cohesion markers (Thompson, 1996). Lexical or grammatical repetitions and conjunctions are markers that can be used to realize the cohesion. A lexical repetition consists in repetition of words. It is a very powerful form of cohesion and it is often used to emphasize or strengthen a concept. As grammatical repetition we consider only the reference, which is. used to indicate whether a term is repeated somewhere earlier in the text (it has already been said) or if it has not yet appeared in the text (it is new). The conjunction is a cohesion marker which connects any two parts of speech. It can be external, when it reflects a state of fact, internal, when it refers exclusively to the organization of the text.

7.6 Analysis and Discussion of the Outcomes

In this section, we analyze the work of one group of students in the study and share the results and outcomes as case studies. In particular, we analyze the students' transcripts with respect to the impact of the social components, (collaboration in task 4 and task 5), and the argumentative component (the ISQ in task 6) on the production of argumentative texts. For this reason, we look at the transcripts related first to the task 2–part 2 and task 3 (individual phase before collaboration and ISQ) and then to task 7 (individual phase after collaboration and ISQ). Moreover, we examine transcripts related to Social Open Question (task 5), Social Logbook (task 9) and Individual Logbook (task 10) to go more in depth.

As was mentioned, a qualitative approach was used to to deepen our understanding of the quality of argumentation. More specifically, for each transcript, we looked at the existence of arguments and the quality (cohesion) of the text. We define the following qualitative model (Table 7.1) which allows us to classify the argumentative texts with respect to the links among various pieces realized by means of cohesion markers (Thompson, 1996).

We assume level 2 as sufficient level of argumentation in the sense that it is perceived that the student has in mind the argument but she is able to communicate only partially. Table 7.2 summarizes the outcomes according to the defined model and level for each student, labelled S#.

As shown in the above data, we can observe a slight change from the task 2 to the task 3. Recall that task 2 part 2 requires the student to reflect on what happens in different cases of her manipulation and task 3 asks for a generalization of what found previously. Only few students, at task 3, are able to do that with partially explicit arguments (mark 2), some others are not able to give any argument (level 0) and nobody is able to write an argumentative text (level 3).

Table 7.1 Qualitative model

Level 0	Text with no argument. An example is S9's answer to task 3: "the angle always remains the same by changing the radius".
Level 1	Text without markers or with generic markers (such as 'and') or with even less generic markers but that do not expose the links. For instance, S8's answer in task 2: "It does not change anything because the angle always remains the same, along with the percentage" (there is a non-generic marker, "because", but the links are not explicit).
Level 2	Some links are also appropriately explained but there are some pieces that are not related: typically, one of the relationships is well highlighted, while the others are in the shadows (that is, it significantly requires the collaboration of the reader). See S8's answer in task 3: "The colored part increases as the circumference and radius increase, but the angle is always the same, i.e. 72° and the percentage is 20%" (there is the explanation of 72° as percentage of 20%, but the speech is not clear and a significant collaboration from the reader is required).
Level 3	All the relevant links are explained through markers. See S8's answer in task 7: "The angle does not vary because it is directly proportional to the circumference. If the radius increases, with it also the circumference, but this does not affect the angle amplitude because grades remain unchanged and also the percentage. The increase of the radius makes the graph clearer when added more data, but the colored part, that is, the red rocks, will always remain 20% and 72°" (non-generic markers, explicit links).

Table 7.2 Levels of argument for each student in task 2, 3 and 7

Task\group	1	2	3	4
2	S1: 1 S2: 2 S3: 0	S4: 0 S5: 0 S6: 1	S7: 0 S8: 1 S9: 0	S10: 1 S11: 1
3	S1: 0 S2: 2 S3: 1	S4: 0 S5: 1 S6: 2	S7: 0 S8: 2 S9: 0	S10: 2 S11: 2
7	S1: 2 S2: 3 S3: 2	S4: 1 S5: 2 S6: 2	S7: 2 S8: 3 S9: 2	S10: 3 S11: 3

Table 7.3 Excerpts from the group 1 in task 2 and task 3

Task	S#	Excerpt	Analysis
2	S1	Nothing changes because the percentages and the degrees are always the same.	Only one external conjunction ("because") not exposing the links.
	S2	Nothing changes because, incrementing the radius, only the circumference grows and thus the degrees are always the same.	Two external conjunctions ("because", "thus"), some pieces are not related.
	S3	The radius increases but the size does not vary.	No argument.
3	S1	The angle does not change varying the radius.	No argument.
	S2	The colored part increases according the increment of the circumference but the angle is constant (72°) as well as the percentage (20%).	Two external conjunctions ("but", "as well as"), more explicit arguments with respect to the one given previously, but some pieces are not related.
	S3	Nothing changes, since the degrees, increasing or decreasing, the circumference remains constant.	Only one external conjunction ("since") not exposing the links.

In the following, we provide detailed analysis of the work of one group; similar results were observed in the other three groups.

We now share some excerpts from the group 1 (Table 7.3). All the students in this group succeeded in IGQ (task 2 part 1), and were then asked if something changes in the circle graph choosing a radius different from the one taken into account by the student during the manipulation (task 2 part 1).

Table 7.2 also shows a meaningful improvement from task 3 to task 7. Let us see some excerpts from the group 1 (Table 7.4). Every student constructs arguments at a sufficient level (level 2), in particular the student S2 improved beyond (from level 2, already reached at task 3, at level 3).

Table 7.4 Excerpts from the group 1 in task 7

Task	S#	Excerpt	Analysis
7	S1	The angle does not vary because it is directly proportional to the circumference, since increasing the circumference the percentage of the angle does not vary, because the degrees are always 360°.	The student goes beyond the sentence constructed by the words-blocks (first part), adding explanations that refer to somehow implicit arguments. Two external conjunctions in the second part ("since", "because") and two lexical repetitions ("angle", "circumference"). Some links are also appropriately explained but there are some pieces not related.
	S2	The angle is directly proportional to the circumference thus it does not vary. Increasing the radius, there is an increase of the size of the circumference but not of the size of the angle within. The degrees of the circumference are always 360° and the percentage is always 100. Then any change we make to the size of the circumference, the angle will remain always constant and fixed.	The student goes beyond the sentence constructed by the words-blocks (first part), adding many explanations that refers to explicit arguments. Two external conjunctions ("but", "then"), some lexical repetitions ("angle", "circumference") and a substitution ("but not of" instead of "but there is not an increase of"). All the relevant links are explained.
	S3	The angle does not vary because it is directly proportional to the area of the circle, because increasing the circumference the angle remains constant. The size of the angle does not change since the radius does not move but it only increases.	The student goes beyond the sentence constructed by the words-blocks (first part), adding explanations that refer to somehow implicit arguments. Two external conjunctions ("since", "because") and a lexical repetition ("angle"). Some links are also appropriately explained but there are some pieces not related.

The tasks 3 and 7 require the student to answer the same question, before and after social tasks (tasks 4 and 5) and ISQ (task 6).

During the discussion before task 5, the students chose to deliver the answer given by S2 (table 2 task 3). They said: "Anyway all the answers are equal". Actually, this is not true (Table 7.2), but they probably refer to the shared belief that the angle does not change. At the same time, the choice of S2's sentence seems to make evident that all the members are somehow implicitly aware that the S2 sentence is the more complete, because it gives an argument too, even if not an explicit argument. We also note that S2 is not the communication official, thus she is not in charge of reporting the shared answer, so this means that the group really choices her sentence.

Next, we examine a chat in task 6:

S3: the explanation?

S1: the reasoning

S3: what do we write?

S2: let me think about

  I guess that the radius is directly proportional to

  I DON'T KNOW:'(((((((((

It is worthwhile to note that this (argumentative component) reveals the fact that an argument is expected and brought the students to think about. It seems not casual that S2 is the one more engaged in the reasoning, as she was the one who already reasoned (implicitly). We note that she seems to be in crisis because she is not able to say in words her thinking but this conflict highlights that something is wrong from the mathematical point of view and she becomes aware of this. In fact, S2 wrote in her logbook:

> I had some difficulties because the size of the circumference could be whatever. And I knew that the answer I constructed with the words-blocks was wrong, but anyway I justified the reasoning after (see Table 7.4).

As conclusion, we share an excerpt from the logbook of the group 1:

> We started the chapter 1 analyzing the give aerogram. After a sequence of procedures justified by an argument, we achieved a common theory, that is: the angle does not vary because it is directly proportional to the area of the circle. The angle is directly proportional to the circumference and thus it does not change. Increasing the radius, there is a growth of the circumference, but not of the size of the angle within. The degrees of the circumference are always 360 and the total percentage is always 100. Thus, any change we make to the size of the circumference, the angle will be always constant and fixed.

The above excerpt highlights a metacognitive process realized by the group working together. As evidenced by the above excerpt from the group's logbook (second line through the end of the excerpt), the group members become aware of a shift from procedures to the need of a justification by arguments and finally to the achievement of a common theory.

Finally, comparing S2's excerpt (Table 7.4) and the above logbook excerpt, we note that S2 acts as an expert in the group and her interactions with the group's mates allow an improvement of each member with respect to the argumentative competence. So, she mediates the underpinning educational objective, as desired, fostered by the script's design, requesting comparisons within the group, shared answer, conversion by words-blocks.

As the script is designed to scaffold argumentation, it was successful, even if the group was not able to give the correct answer from the mathematical point of view. Looking at the transcripts, students achieved an intuitive understanding of the mathematics but they lack some contents (i.e. direct proportionality) needed to convert their understanding in texts.

7.7 Conclusions

In this paper, we have introduced a general methodology to support competence-based learning. This method, DIST-M, represents an adaptation of the DIST methods, specifically created to support mathematics learning in an e-learning environment. The specific activity discussed focused on developing argumentative competence, using DIST-M methodology within a Moodle platform. The implementation involved using computer-supported collaboration scripts aimed to foster the students' shift from investigating and reasoning to communicate verbally what found, by constructing sentences with evidence of arguments in a style typical of scientific communication. The implementation utilized the Moodle features and supported collaboration among peers as well as individual work. Collaborative tasks utilize the chat or forum features, depending on what linguistic register is to be supported, and the particular setting of the forum to be used (Questions and Answer) avoid the students undue influence on each other. Analogously, the group features in Moodle-hosted tasks compels the students to take part in the activity. Individual tasks also benefit from the technology: the student can manipulate dynamical objects, graphical or textual, with automatic tracking, thus personalized recovery paths are delivered.

The analysis of the students' transcripts in chat and forum, in a linguistic perspective, have shown an improvement on the level of arguments given by each student. Moreover, they show how the script supports reasoning on mathematical concepts.

Looking at the tasks' flow, we conjecture that the improvement has been mainly fostered by two key points:

- the social tasks, that require to negotiate a delivery shared by all the members of the group, seem to come to light the need of producing arguments to support the answer; this request was present from the beginning, but many students did not give arguments when delivered individually;
- the argumentative task, that supports the treatment from the sentence in a colloquial register chosen by the group to sentence in a literate register, foster not only to refine the argument, but mainly to deepen the students' mathematical understanding as shown by their further cohesive sentences produced to explain their reasoning.

The above outcomes encourage further investigation of the design and effectiveness of DIST-M for promoting the ability of converting reasoning in constructing arguments expressed by cohesive texts.

References

Albano, G., Dello Iacono, U., & Fiorentino, G. (2016). An online Vygotskian learning activity model in mathematics. *Journal of e-Learning and Knowledge Society, 12*(3), 159–169.

Albano, G., Dello Iacono, U., & Mariotti, M. A. (2017). A computer-based collaboration script to mediate verbal argumentation in mathematics. In Dooley, T. & Gueudet, G. (Eds.). *Proceedings of the Tenth Congress of the European Society for Research in Mathematics Education* (CERME10, February 1–5, 2017). Dublin, Ireland: DCU Institute of Education & ERME.

Andriessen, J., Baker, M., & Suthers, D. (2003). Argumentation, computer support, and the educational context of confronting cognitions. In J. Andriessen, M. Baker, & D. Suthers (Eds.), *Arguing to learn: Confronting cognitions in computer-supported collaborative learning environments* (pp. 1–25). Dordrecht: Kluwer.

Baker, M. (2003). Computer-mediated argumentative interactions for the co-elaboration of scientific notations. In J. Andriessen, M. Baker, & D. Suthers (Eds.), *Arguing to learn: Confronting cognitions in computer-supported collaborative learning environments* (pp. 47–78). Dordrecht, Netherlands: Kluwer.

Boero, P. (1999). Argumentation and mathematical proof: A complex, productive, unavoidable relationship in mathematics and mathematics education. International Newsletter on the Teaching and Learning of Mathematical Proof (July/August 1999).

Chi, M. T. H., Bassok, M., Lewis, M. W., Reimann, P., & Glaser, R. (1989). Self-explanations: How students study and use examples in learning to solve problems. *Cognitive Science, 13*, 145–182.

Ferrari, P. L. (2004). Mathematical language and advanced mathematics learning. In M. Johnsen Høines & F. A. Berit (Eds.), *Proceedings of the 28th Conference of the International Group for the Psychology of Mathematics Education: PME 28* (pp. 383–390). University College Bergen (N).

Halliday, M. A., & Hasan, R. (1976). *Cohesion in english*. London: Longman.

King, A. (2007). Scripting collaborative learning processes: A cognitive perspective. In F. Fischer, I. Kollar, H. Mandl, & J. Haake (Eds.), *Scripting computer-supported collaborative learning: Cognitive, computational and educational perspectives* (pp. 13–37). New York: Springer.

Kollar, I., Ufer, S., Reichersdorfer, E., Vogel, F., Fischer, F., & Reiss, K. (2014). Effects of collaboration scripts and heuristic worked examples on the acquisition of mathematical argumentation skills of teacher students with different levels of prior achievement. *Learning and Instruction, 32*, 22–36.

Kuhn, D., Shaw, V., & Felton, M. (1997). Effects of dyadic interaction on argumentative reasoning. *Cognition and Instruction, 15*(3), 287–315.

Laurillard, D. (2013). *Teaching as a design science: Building pedagogical patterns for learning and technology*. London: Routledge.

Mariotti, M. A. (2006). Proof and proving in mathematics education. In A. Gutiérrez & P. Boero (Eds.), *Handbook of research on the psychology of mathematics education: Past, present and future* (pp. 173–204). Rotterdam: Sense.

Pea, R. D. (1994). Seeing what we build together: Distributed multimedia learning environments for transformative communications. Special Issue: Computer support for collaborative learning. *Journal of the Learning Sciences, 3*(3), 285–299.

Sfard, A. (2001). Learning mathematics as developing a discourse. In R. Speiser, C. A. Maher, & C. N. Walter (Eds.), *Proceedings of the Twenty-Third Annual Meeting of the North American Chapter of the International Group for the Psychology of Mathematics Education* (pp. 23–43). Columbus, OH: ERIC Clearinghouse for Science, Mathematics, and Environmental Education.

Thompson, G. (1996). *Introducing functional grammar*. London: Routledge.

Turner, R., & Adams, R. J. (2012). *Some drivers of test item difficulty in mathematics: An analysis of the competency rubric*. Paper presented at AERA Annual Meeting, April 2012, Vancouver, Canada.

Vogel, F., Kollar, I., Ufer, S., Reichersdorfer, E., Reiss, K., & Fischer, F. (2015). Fostering argumentation skills in mathematics with adaptable collaboration scripts: only viable for good self-regulators? In O. Lindwall, P. Häkkinen, T. Koschmann, P. Tchounikine, & S. Ludvigsen (Eds.), *Exploring the material conditions of learning. The Computer Supported Collaborative Learning Conference (CSCL) 2015—Volume II* (pp. 576–580). Gothenburg: International Society of the Learning Sciences, University of Gothenburg.

Vygotsky, L. S. (1978). *Mind in society: The development of higher psychological processes.* Cambridge, MA: Harvard University Press.

Weinberger, A., Kollar, I., Dimitriadis, Y., Mäkitalo-Siegl, K., & Fischer, F. (2009). Computer-supported collaboration scripts: Perspectives from educational psychology and computer science. In N. Balachef, S. Ludvigsen, T. De Jong, A. Lazonder, & S. Barnes (Eds.), *Technology-enhanced learning: Principles and products* (pp. 155–174). Dordrecht: Springer.

Weinberger, A., Stegmann, K., Fischer, F., & Mandl, H. (2007). Scripting argumentative knowledge construction in computer-supported learning environments. In F. Fischer, H. Mandl, J. Haake, & I. Kollar (Eds.), *Scripting computer-supported communication of knowledge-cognitive, computational and educational perspectives* (pp. 191–211). New York: Springer.

Zan, R. (2011). The crucial role of narrative thought in understanding story problems. In K. Kislenko (Ed.): *Current state of research on mathematical beliefs XVI* (pp. 287–305). Tallinn Estonia: Tallinn University.

Chapter 8
Effective Use of Math E-Learning with Questions Specification

Yasuyuki Nakamura, Kentaro Yoshitomi, Mitsuru Kawazoe, Tetsuo Fukui, Shizuka Shirai, Takahiro Nakahara, Katsuya Kato and Tetsuya Taniguchi

Abstract MATH ON WEB, STACK, and Maple T.A. are the prominent mathematics e-learning systems used in Japan. They can assess answers containing mathematical content freely written by students as opposed to only answers to multiple-choice questions. However, there are two major challenges while using these systems: inconvenience in inputting answers and heavy content-development workload. We have developed two math input interfaces, MathTOUCH and FlickMath, using which students can easily input mathematical expressions. The interfaces were developed as part of a project aimed at accelerating the spread of math e-learning systems using a question-sharing environment among heterogeneous systems such as MATH ON WEB and Maple T.A. Further, they form a part of mathematics e-learning question specification ('MeLQS') system, which is currently being developed in our project to realise this objective. We would like to emphasize the importance of building a common base, 'MeLQS', for creating questions in math e-learning.

Y. Nakamura (✉)
Graduate School of Informatics, Nagoya University,
A4-2, 780, Furo-cho, Chikusa-ward, Nagoya 464-8601, Japan
e-mail: nakamura@nagoya-u.jp

K. Yoshitomi · M. Kawazoe
Osaka Prefecture University, Sakai, Japan

T. Fukui · S. Shirai
Mukogawa Women's University, Nishinomiya, Japan

T. Nakahara
Sangensha LLC., Chitose, Japan

K. Kato
Cybernet Systems Co., Ltd., Tokyo, Japan

T. Taniguchi
Nihon University, Tokyo, Japan

Present Address:
S. Shirai
Osaka University, Toyonaka, Japan

© Springer International Publishing AG, part of Springer Nature 2018
J. Silverman and V. Hoyos (eds.), *Distance Learning, E-Learning and Blended Learning in Mathematics Education*, ICME-13 Monographs,
https://doi.org/10.1007/978-3-319-90790-1_8

133

Keywords STACK · MATH ON WEB · Maple T.A. · Math e-learning
Question sharing

8.1   Introduction

In recent years, information and communication technology infrastructure has
improved in schools and e-learning has become increasingly popular. One of the
most important functions of e-learning is automatic assessment to evaluate a student's understanding of course content. Computer-aided assessment (CAA) is an
old technique and was applied to many subjects, even before learning management
systems (LMS) became popular. One of the most common question types in CAA
systems is multiple-choice questions (MCQ), wherein the potential answers are
provided by a teacher and students select a single response as their answer.
A well-constructed MCQ provides a correct answer with plausible distracters,
which are usually decided by knowledge of common student errors.

CAA has been successfully used in language education. For example, CAA has
been applied to placement tests based on Item Response Theory and computer
adaptive testing has been compared with pencil and paper testing in terms of
validity and efficiency (Koyama & Akiyama, 2009).

CAA has also been carried out using MCQs in scientific subjects, but the MCQ
format is not sufficient to evaluate a student's comprehension level. For example,
when students answer MCQ type questions, students can simply choose an answer
from a list even if they do not know the correct answer, and there is a possibility of
answering correctly by guessing. In order to avoid these problems with MCQs, it is
preferable to adopt a question format wherein students provide answers containing
mathematical expressions, which are subsequently evaluated. There are some systems that evaluate student-provided answers. In one of the major systems,
CIST-Solomon (CIST, 2016; Komatsugawa, 2004), which has more than 30,000
subject areas including mathematics, students construct mathematical expressions
using software such as Flash.

In this study, we focus on a CAA system where students provide mathematical
expressions through keyboard input and the answer is evaluated using a computer
algebra   system   (CAS).   For   example,   for   the   differentiation   question
$\frac{d}{dx}\left(\frac{1}{4}x^2 + \frac{1}{2}x + 1\right)$, the correct answer is $\frac{1}{2}(x+1)$, but some students provide $\frac{1}{2}x + \frac{1}{2}$,
others provide $\frac{x+1}{2}$, etc. These answers are all mathematically equivalent and cor-
rect, and the evaluation of equivalence is underpinned by CAS. Recently, this kind
of math e-learning system has become popular and, to the best of our knowledge,
there are three main systems being used in Japan: MATH ON WEB, STACK, and
Maple T.A. Although the importance of questions with student-provided answers is
understood in e-learning of mathematics, many teachers are now seeking effective
ways to carry out mathematics e-learning using CAA. In this paper, we summarize
the utilization of math e-learning systems in Japan, present some associated

problems, and propose solutions to a variety of issues with implementing these technologies.

8.2 Math E-Learning Systems in Japan

Hereafter, we focus on online assessment systems for mathematics using CAS as a mathematics e-learning system. In contrast to multiple-choice or true-or-false choice questions, which are referred to as teacher-provided-answer questions, online assessment for mathematics provides student-provided-answer questions, wherein students submit numerical values or mathematical expressions as answers to the questions. In order to assess answers to student-provided-answer questions, math e-learning systems evaluate these answers using CAS. In this section, three math e-learning systems are briefly discussed: MATH ON WEB, STACK, and Maple T.A. These math e-learning systems, which are representative of the systems in use in Japan, use *Mathematica*, Maxima, and Maple, respectively, as their CAS to evaluate student answers. The problems associated with the continued use of these systems are outlined at the end of this section.

8.2.1 MATH ON WEB

E-learning/e-assessment systems based on *webMathematica* have been developed and are currently being used in mathematics education for first-year students at Osaka Prefecture University (2016; Kawazoe, Takahashi, & Yoshitomi, 2013; Kawazoe & Yoshitomi, 2016a, b). The systems are available on the 'MATH ON WEB' website. The website has two systems: web-based mathematics learning system (WMLS) and web-based assessment system of mathematics (WASM).

WMLS is a self-learning system aimed at promoting students' after-class learning. It has two sections: drill section and simulation section. The drill section offers an online mathematics exercise environment with more than 1000 mathematics problems in calculus and linear algebra courses for first-year university students. When a student submits his/her answer to a presented problem, the system analyses the answer using *Mathematica* and provides a feedback message. When the answer is incorrect, the system provides a different feedback message according to the error type identified. The simulation section offers simulation type content that assists students in learning mathematical concepts.

WASM is an online mathematics assessment system with two different modes: assessment mode and exercise mode. The basic mechanism of WASM is the same as the drill mode of WMLS, but in WASM, problems are presented in random order or are randomly generated by *Mathematica*. In the assessment mode, online assessment tests are classified with respect to learning units and achievement levels, and students can assess their achievement in each learning unit using the online test associated with the learning unit. In the exercise mode, students can carry out problem-solving exercises in each assessment test. WASM has various improvements over WMLS. One such improvement is the implementation of popup keyboards based on jQuery Keyboard (Wood, 2014), which enables students to input their answers (Kawazoe & Yoshitomi, 2016b).

At Osaka Prefecture University, more than 600 students use the systems annually and both systems are used to promote students' after-class learning and the implementation of blended learning environments (Kawazoe & Yoshitomi, 2016a). Quantitative analysis of the log data in WMLS (Kawazoe et al., 2013) showed that students use the system mainly in the late afternoons and at nights. Hence, it can be concluded that WMLS promotes students' after-class engagement. Statistical analysis of engineering students (ibid.) shows that there is a positive correlation between the frequency of use of the system and achievement in mathematics. Kawazoe and Yoshitomi (2016b) reported on a blended learning linear algebra class with WMLS and WASM for first-year engineering students and noted that many students stated that the blended learning approach is useful and preferable.

The next objective of the MATH ON WEB project is to develop a Moodle plugin for WMLS and WASM. The development of the Moodle plugin is still underway, but a prototype has already been developed (Nakahara, Yoshitomi, & Kawazoe, 2016).

8.2.2 STACK

STACK, developed by Sangwin (2013), uses Maxima as its CAS to evaluate students' answers. STACK not only assesses the mathematical equivalence of students' answers but also generates outcomes, such as providing feedback according to the mathematical properties of students' answers. The feedback function is implemented using the potential response tree (PRT) mechanism. PRT is an algorithm that establishes the mathematical properties of students' answers and provides feedback specifically to each student.

STACK is being utilised for several subjects in many institutions in Japan. Taniguchi, Udagawa, Nakamura, and Nakahara (2015), who used STACK in a math class at Nihon University, used logistic regression analyses to document that STACK is effective. STACK is also being used in physics classes at Toyama University and Nagoya University. Basic support for scientific units has been added to the latest version of STACK; this function is expected to enhance the use of STACK in physics class.

STACK is an open source system and users can develop required functions. For example, the plot function in STACK is poor. Specifically, only the drawing of single variable functions is supported. The plot function has been enhanced using Maple (Nakamura, Amano, & Nakahara, 2011) and gnuplot (Fukazawa & Nakamura, 2016).

8.2.3   Maple T.A.

Maple T.A. is a web-based online testing and assessment system developed by Maplesoft, a Canadian software company. It was designed especially for science, technology, engineering, and mathematics (STEM) courses. Further, it offers various question types, flexible assignment properties, full-featured gradebook with reporting and analytical tools, seamless connectivity to any LMS, and support for multiple languages. In recent years, Maple T.A. has been gradually and steadily adopted for STEM education in academic institutions such as high schools, colleges, and universities in Canada, U.S.A., many European countries, China, and Taiwan (Maplesoft, 2016). After adopting and utilizing Maple T.A., they not only successfully reduced their grading burden but also improved the STEM education learning environment, resulting in students being strongly engaged in the courses. Consequently, they also view Maple T.A. as an important teaching tool.

By contrast, in Japan, Maple T.A. has been promoted by its distributor Cybernet Systems, a Japanese software company, for several years and is gradually beginning to be recognized as an online testing and assessment system for STEM education. Cybernet Systems already has a user—the Faculty of Science and Technology at Ryukoku University. The faculty utilizes Maple T.A. for fundamental mathematics education such as pre-entrance education and remedial education, bridging the gap between high schools and university. They state that Maple T.A. is capable of improving and enhancing the basic math ability of students and thus, they have plans to expand the use of Maple T.A. to a wide variety of STEM subjects in the future (Higuchi, 2015).

Cybernet Systems experimentally introduced Maple T.A. in a STEM programming course at Gakushuin University, and also assessed the functionality and performance of Maple T.A. (Cybernet Systems, 2012). The course provided students with a small test designed specifically to measure the programming skills of Maple. Eventually, all the results associated with this course are managed within Maple T.A. along with all the results of external items such as offline assignments graded using a Maple T.A. rubric.

In 2016, Cybernet Systems launched and conducted a project aimed at evaluating the capabilities and performance of Maple T.A. in seminars and classes with students, and obtained the cooperation of math instructors at universities. Six math instructors from different universities joined the project and evaluated Maple T.A. in their own math seminars and classes, with capacity ranging from several students to more than 200 students. Their main objective was to ascertain whether Maple T.A. enabled them to efficiently streamline the existing grading workflow and reduce the workload associated with creating questions, marking tests, and managing gradebooks. The instructors were satisfied with Maple T.A to some extent. Furthermore, Cybernet Systems obtained diverse feedback from the instructors and students, which centered around two major issues. One was a need to expand the math content from the viewpoint of the instructors; the other was a need to improve the math input interface from the viewpoint of the students.

8.2.4   Common Challenge

E-learning systems, especially CAS-based mathematics e-learning systems, are undoubtedly powerful tools for developing effective mathematics learning environments. However, CAS-based e-learning systems, including the aforementioned three systems, have two issues in common: inconvenience in inputting answers and heavy workload in developing content.

In CAS-based mathematics e-learning systems, students have to input their answers using the input form provided by CAS, which students consider inconvenient. Such inconvenience should be resolved; hence, developing effective math input interfaces for these systems is important. We discuss this issue in detail in Sect. 8.3.

The issue of heavy content-development workload originates from the fact that, in many Japanese universities, e-learning content is usually developed by teachers —specifically, a very limited number of teachers. If the teachers could share the content or resources beyond system differences, their burdens would be reduced. We investigate this problem in Sect. 8.4 and we will determine the importance of building a common base for creating questions in Sect. 8.5.

If the aforementioned two issues can be overcome, the use of e-learning systems in university mathematics education would become more pervasive.

8.3   Math Input Interfaces

As described at the end of the previous section, one of the problems associated with math e-learning systems is the math input complexity for questions requiring the input of mathematical expressions as answers rather than multiple-selection or number input types of answers. For example, in order to input $3x^2 - \frac{2x}{(x^2+1)^2}$, which is an answer for the differentiation $\frac{d}{dx}\left(x^3 + \frac{1}{x^2+1}\right)$, the answer should be "3*x^2-2*x/(x^2+1)^2", which easily causes typing errors. It has been reported that many students experienced syntax issues in answering questions (Thaule, 2016). Several interfaces aiming to minimize math input difficulties have been proposed. The proposed interfaces include DragMath (Sangwin, 2012), a drag and drop equation editor in the form of a Java applet, which is used as one of the math input types for STACK. However, the interface requires Java and the input environment is restricted.

Most math e-assessment systems use CAS (e.g. *Mathematica*, Maple, Maxima etc.), which requests users to input mathematical expressions according to the rule of CAS. In order to improve the convenience of math input, template-based interfaces, such as DragMath mentioned above, are added to text-based interfaces. However, it is difficult for novice learners to adapt to current standard interfaces. For example, text-based interfaces accept input according to the CAS command syntax and it is difficult for the users to imagine the desired mathematical expressions because the input is not in the WYSIWYG format. However, structure-based interfaces have an advantage in that learners can operate in the WYSIWYG format. Moreover, they can input using math template icons. Therefore, they do not need to remember CAS command syntax in the case of the text-based interface. However, learners should understand the structure of the required mathematical expressions and should be able to select the math template icons in the correct order (Pollanen, Wisniewski, & Yu, 2007). Furthermore, it is cumbersome to make corrections later (Smithies, Novins, & Arvo, 2001).

In this section, we present two math input interfaces, MathTOUCH and FlickMath, developed by the authors. MathTOUCH accepts math input in the form of colloquial-style mathematical text and FlickMath supports math input on mobile devices.

8.3.1   MathTOUCH

MathTOUCH is a mathematical input interface with Java that facilitates conversion from colloquial-style mathematical text (Fukui, 2012). With this interface, users do not need to enter symbols that are not printed. For example, if users would like to enter $\frac{x^2+1}{3}$, they have only to enter "x2+1/3". They do not need to input parentheses for the delimiters, and a power sign (e.g. a caret symbol) as a list of candidates for each mathematical element is shown in WYSIWYG based on the user input (see Fig. 8.1). After all the elements are chosen interactively, the desired mathematical expression can be created.

In a previous study, Shirai and Fukui implemented MathTOUCH in STACK and conducted two experiments—a performance survey (Shirai, Nakamura, & Fukui, 2015) and an eight-week learning experiment (Shirai & Fukui, 2014)—to evaluate the efficacy of MathTOUCH. The results obtained indicated that MathTOUCH enables tasks to be completed approximately 1.2–1.6 times faster than standard input interfaces (such as text-based interfaces and structure-based interfaces).

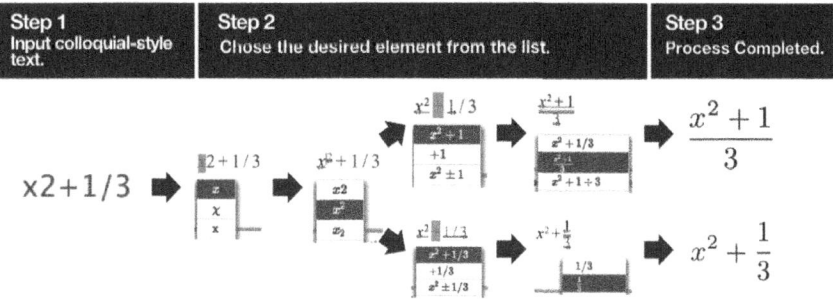

Fig. 8.1 MathTOUCH input procedure

Moreover, MathTOUCH was shown to have a high level of satisfaction with respect to math input usability. The results of the eight-week learning experiment show that students could practice using MathTOUCH on STACK at the same learning rate as with the standard input interface on STACK.

In 2016, Shirai and Fukui reconstructed MathTOUCH using JavaScript to make MathTOUCH available not only on Java-compliant devices but also on various other devices, and conducted a five-week learning experiment to evaluate the stability of reconstructed MathTOUCH. The results showed that students can study using reconstructed MathTOUCH on STACK as effectively as with the previous Java-version of MathTOUCH. The details including the data are available in Shirai and Fukui (2017).

8.3.2   MathDox and FlickMath

Nakamura, Inagaki, and Nakahara (2014a) developed an input interface for STACK using MathDox formula editor. MathDox formula editor facilitates the input of mathematical formulas two-dimensionally using a keyboard, and it also has a palette for input assistance (Fig. 8.2). Maple T.A.'s Equation Editor also realises the same type of mathematical expressions. However, with both editors, users have to switch keyboard between letters and numbers/symbols, especially when using mobile devices. Further, the editors do not reduce the complexity of the math input process for such devices.

Based on the MathDox input type, a new input type, FlickMath, was developed for using STACK on mobile devices (Nakamura & Nakahara, 2016). FlickMath allows the input of mathematical expressions by the flick operation (Fig. 8.3). The flick operation is carried out by placing a finger on the prepared keyboard, shifting the finger vertically or laterally, and subsequently releasing it. Japanese students often use flick input to input characters on their smartphones; therefore, inputting math using the flick operation should be natural for them. On tablet devices, a full

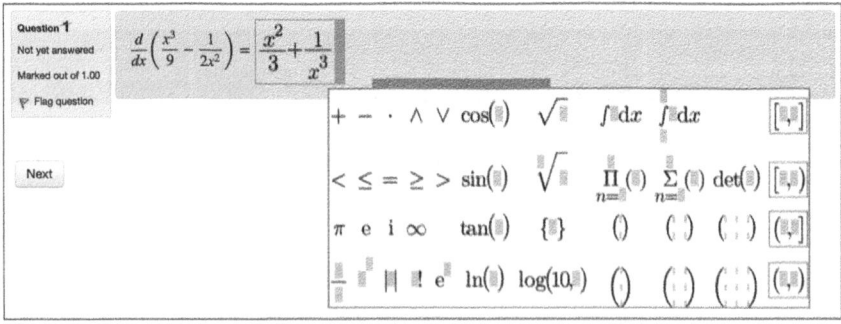

Fig. 8.2 MathDox input type for STACK

Fig. 8.3 FlickMath input type for STACK

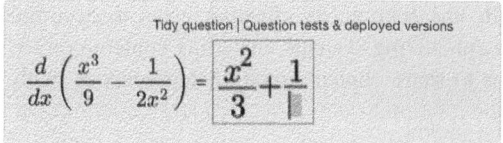

keyboard is displayed and the flick operation is implemented. We conducted a survey on the usability and satisfaction levels of students. Our results, based on the average responses of 29 students, indicate that usability and satisfaction levels are higher when the flick input method is used to enter mathematical expressions as compared to the direct input method (Nakamura & Nakahara, 2017). As the input operation using FlickMath is less cumbersome than using the direct method on a smartphone, it could be effective for drill practices on mobile devices when students have spare time.

8.4   Sharing Questions

When e-learning is employed, not only in mathematical science but also other subjects, content has to be prepared. However, math e-learning, especially math online testing, has a relatively short history and an individual teacher may not have sufficient content. As stated in Sect. 8.2.4, developing questions that can assess students' answers and return suitable feedback appropriately is time-consuming. Accordingly, a question-authoring tool that reduces the amount of work involved was developed for STACK (Nakamura, Ohmata, & Nakahara, 2012). It is helpful for authors; however, in order to promote math e-learning effectively, it is perhaps more important to prepare as much high-quality content as possible and to share such content among teachers.

In this section, as effective ways to accumulate content, we discuss several content-sharing systems, including content-converting methods and our project that aims to share content among heterogeneous math e-learning systems.

8.4.1 Item Bank System: MathBank

In order to effectively promote student learning, well-structured questions must be shared, thereby allowing any registered user to access them. This helps to broaden the applications of e-learning to math and science subjects. Accordingly, we developed an item bank system (Nakamura, Taniguchi, & Nakahara, 2014b), called 'MathBank', open to the public at Mathbank.jp, https://mathbank.jp/.

We developed MathBank as a Moodle system wherein any user can register questions and search for registered questions after user authentication. When users register a question in MathBank, they are prompted to include metadata associated with the question such as grade, difficulty level, publicity level, and keywords. Questions can be registered by uploading an XML file via the interface. Users can also create questions on MathBank itself. After searching through the list of questions, users can download STACK questions in XML format and import the file to their servers for subsequent use. MathBank also provides time for registered questions to be tested straight from MathBank, which creates stored logs. The log is used to reconsider the difficulty level and to improve the quality of the questions in the system. MathBank was opened to the public approximately three years ago and, as of December 2016, 46 users were registered and approximately 200 questions were available on the system.

8.4.2 Maple T.A. Cloud

Tens of thousands of questions are available in the Maple T.A. Cloud (Maplesoft, 2013), a worldwide content-sharing system based on Maple T.A., ready for immediate use or customization. Content is available for a variety of subjects, including calculus, pre-calculus, algebra, differential equations, linear algebra, physics, chemistry, engineering, statistics, and economics.

8.4.3 Converting Content Between Different Systems and Building Common Base for Content Creation

The aforementioned question-sharing systems, Mathbank and Maple T.A. Cloud, are designed for specific math e-learning systems STACK and Maple T.A., respectively. One can share questions in each system only. A solution for increasing the amount of content is converting high-quality content created in one system to another system. There are some projects to convert questions among different math e-learning systems; e.g. conversion system between Maple T.A. and SOWISO (Baumgarten et al., 2015) and Maple T.A. to STACK conversion (Higuchi, 2016). However, conversion is not always perfect since some features of a system are sometimes not supported by the other system.

8.4.4   Necessity of Common Base for Sharing Content

We reviewed two content-sharing systems and some content-converting methods. However, it is certainly preferable to have a common base of shared content to accumulate content. An aggregation of content can be compared with an emerging "system" of interactive elements. For example, knowledge map (Chen, Chu, Chen, & Su, 2016; Reed, 2013) is a visualization of connections among subjects and shows an interaction between elements, which is similar to content-sharing systems. Further, the elements of content-sharing systems should be increased in number and, in other words, content-sharing systems should emerge. It is said that "evolutionary algorithms have a common base with evolution, since they are based on fundamentals of natural selection" (Davendra, 2010). Our project aims to share content among heterogeneous math e-learning systems based on a common base, i.e. mathematics e-learning question specification (MeLQS), which we believe realizes the objective of this work, i.e. to determine effective ways to carry out mathematics e-learning using CAA.

8.5   Mathematics E-Learning Questions Specification: MeLQS

In order to build a common base for sharing questions, we verified that the structures of the question data in STACK and WASM (see Sect. 8.2.1) are analogous (Yoshitomi & Kawazoe, 2013). In WASM, the data consists of question sentences with formatting similar to HTML that permits MathJax description, which has a similarly formatted answer format; *Mathematica* to analyze the input; a set of program (maybe including random functions) parameters that are randomly selected when actually used; feedback messages coupled with the return code of the analysis program above. It is well known that STACK has a similar constitution; conceptually, both the systems have the same structure. We attempted to convert the question data between the systems manually (Yoshitomi, 2014). First, we converted each piece of data to better understand text describing what the author wants to do and how the parameters are determined. We called the prototype of specification of the question data MeLCS, Mathematics e-Learning Contents Specification, (Yoshitomi, 2014), but have since changed the name to MeLQS, as defined previously.

We started the MeLQS project, a four-year project, with grant support in 2016, and aim to share the e-learning/e-assessment content in the universal format MeLQS. This format is expected to be easily exported to any format available in the world, including MATH ON WEB, STACK, and Maple T.A. Therefore, we use MeLQS as a common base of shared content to accumulate questions for math e-learning. After the preliminary analysis of structures of the questions of MATH ON WEB and STACK, we determined it appropriate to categorize the structures of questions as follows: question text and routine to create it; definition of answer column and answer type; routine to evaluate student answer and feedback. MeLQS also has metadata of questions: question name, subject, intention behind a question etc.

One of the most important features of MeLQS is that it is constructed with two specifications: concept design and implementation specification. In the following subsection, we describe MeLQS in detail.

8.5.1 Concept Design for Questions

Concept design is a specification of questions that describes how the question is designed according to concept, and it is described by mathematical statements rather than programming statements so that all mathematics teachers can understand the concept addressed by the question. The concept design is stored in a database and databased concept design can be viewed on the MeLQS web system or can be exported to TeX and PDF format. Therefore, concept design is useful not only for online tests but also paper-based tests. We implemented an authoring tool of concept design as a Moodle plug-in that allows users to, create a concept design, including metadata: question name, subject, intention behind a question etc., step-by-step as shown in Fig. 8.4. At the present stage, how to input mathematical expressions is not fixed, but TeX is preferable for the preview. We also plan to familiarize the editorial function to all teachers with support from MathTOUCH (see Sect. 8.3.1).

8.5.2 Implementation Specification for Questions

The standard of implementation specification is being formulated as of May 2017. It is considered that those who have experience in authoring questions for online tests would create implementation specification based on the suggested concept design. In the implementation specification, details of settings of questions defined as dependencies on each math e-learning system are eliminated. For example,

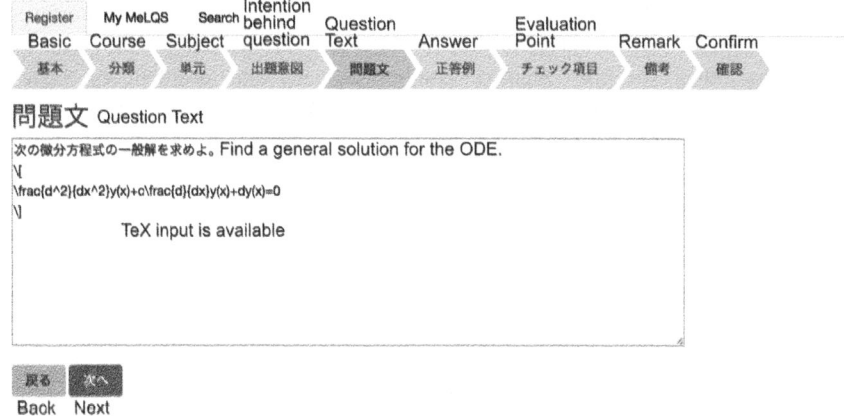

Fig. 8.4 Authoring tool of concept design of MeLQS

input of mathematical expression should not be dependent on each CAS syntax. Authoring tool like Fig. 8.4 will be developed in the future.

8.5.3 Implementation of Questions in Math E-Learning System

Questions based on the implementation specification can be used in each mathematics e-learning system. We plan to provide MeLQS as a cloud service with functions that enable users to author the question data and to export and import them to heterogeneous systems. At the present stage, the evaluation procedure implemented in STACK and MATH ON WEB cannot be reflected to Maple T.A. but the implementation is being considered by referring MeLQS.

In the future, we aim to make it easy for virtually all teachers to participate in the service. All the users can use this system freely but are expected to provide feedback about the effects or issues associated with the downloaded teaching materials to the community. We expect that significant use of the service will promote more practical usage and increase the efficacy of math e-learning systems.

8.6 Conclusion

We briefly reviewed the three main math e-learning systems used in Japan and outlined two problems associated with using them: inconvenience in inputting answers and heavy content-development workload. In order to solve the problem of inconvenience in inputting mathematical expressions, we developed two math input interfaces: MathTOUCH and FlickMath. Currently, both the math input interfaces are implemented only in STACK, but as they are HTML5-based, they can be applied to other systems. Sharing content reduces the problem of heavy content-development workload. Maple T.A. Cloud and MathBank are used to share mathematical questions by Maple T.A. and STACK, respectively. Although similar types of mathematical questions are present in both the systems, these questions cannot always be interchanged between them. Many questions have also accumulated in MATH ON WEB, but these questions are not compatible with other systems.

Collecting well-constructed mathematical questions in the promotion of math e-learning is undoubtedly important. In order to share questions among heterogeneous math e-learning systems—MATH ON WEB, STACK, and Maple T.A.—we have started a four-year project wherein the first step is to design the universal format MeLQS. We believe MeLQS, as a common base, is necessary for contents and usage of e-Learning systems in undergraduate mathematics education to increase in number. This is our answer to the research question in the present paper: what is necessary to realize effective use of mathematics e-learning using CAA. Eventually, we plan to provide MeLQS cloud service with functions that enable users to author question data and to export and import them to heterogeneous systems. We believe that heavy use of the service will promote more practical usage and increase the efficacy of math e-learning systems.

Acknowledgements This work was supported by JSPS KAKENHI Grant Number 16H03067.

References

Baumgarten, K., Brouwer, N., Cohen, M., Droop, B., Habbema, M., Heck, A., et al. (2015). *Design and implementation of a conversion method between mathematical question bank items.* http://starfish.innovatievooronderwijs.nl/media/uploads/MathConverter_report_def.pdf. Accessed July 12, 2017.

Chen, T.-Y., Chu, H.-C., Chen, Y.-M., & Su, K.-C. (2016). Ontology-based adaptive dynamic e-learning map planning method for conceptual knowledge learning. *International Journal of Web-Based Learning and Teaching Technologies (IJWLTT), 11*(1), 1–20.

Chitose Institute of Science and Technology. (2016). CIST-Solomon, https://solomon.mc.chitose.ac.jp/CIST-Shiva/. Accessed July 12, 2017.

Cybernet Systems. (2012). A case study at Gakusyuin University. Online: http://www.cybernet.co.jp/maple/documents/pdf/mac2012doc/MAC2012-1-1.pdf.

Davendra, D. (2010). Evolutionary algorithms and the edge of chaos. In: I. Zelinka, S. Celikovsky, H. Richter, & G. Chen (Eds.), *Evolutionary algorithms and chaotic systems.* Studies in Computational Intelligence (Vol. 267). Berlin: Springer.

Fukazawa, K., & Nakamura, Y. (2016). Enhancement of plotting environment of STACK with Gnuplot. *JsiSE Research Report, 31*(3), 6–9 (in Japanese).

Fukui, T. (2012). An intelligent method of interactive user interface for digitalized mathematical expressions. *RIMS Kokyuroku, 1780,* 160–171 (in Japanese).

Higuchi, S. (2015). Study of mathematical software and its effective use for mathematics education. *RIMS Kokyuroku, 1978,* 72–78 (in Japanese).

Higuchi, S. (2016). Maple T.A. to STACK conversion, https://github.com/hig3/mta2stack. Accessed July 12, 2017.

Kawazoe, M., Takahashi, T., & Yoshitomi, K. (2013). Web-based system for after-class learning in college mathematics via computer algebra system. In M. Inprasitha (Ed.), *Proceedings of the 6th East Asia Regional Conference on Mathematics Education* (Vol. 2, pp. 476–485). Phuket, Thailand: Khon Kaen University.

Kawazoe, M., & Yoshitomi, K. (2016a). Construction and operation of mathematical learning support web-site MATH ON WEB developed by webMathematica. *Bulletin of JSSAC, 22*(1), 13–27 (in Japanese).

Kawazoe, M., & Yoshitomi, K. (2016b). E-learning/e-assessment systems based on webMathematica for university mathematics education. *Submitted to MSOR Connections.*

Komatsugawa, H. (2004). Development of e-learning system using mathematical knowledge database for remedial study. In *Proceedings of International Conference on Computers and Advanced Technology in Education* (pp. 212–217).

Koyama, Y., & Akiyama, M. (2009). Developing a computer adaptive ESP placement test using moodle. In *Proceedings of E-Learn: World Conference on E-Learning in Corporate, Government, Healthcare, and Higher Education* (pp. 940–945).

Maplesoft. (2013). Maple T.A. Cloud facilitates new ways for content collaboration and sharing between users, http://www.maplesoft.com/company/news/releases/2013/2013-01-15-Maple-TA-Cloud-facilitates-new-ways-for-.aspx. Accessed July 14, 2017.

Maplesoft. (2016). Maple T.A. user case studies. Online: http://www.maplesoft.com/company/casestudies/product/Maple-TA/. Accessed July 12, 2017.

Nakahara, T., Yoshitomi, K., & Kawazoe, M. (2016). Development of a Mathematica-based Moodle plugin for assessment in mathematics. *JsiSE Research Report, 31*(2), 15–16 (in Japanese).

Nakamura, Y., Amano, H., & Nakahara, T. (2011). Enhancement of plotting function of math e-learning system STACK. In *Proceedings of E-Learn: World Conference on E-Learning in Corporate, Government, Healthcare, and Higher Education 2011* (pp. 2621–2627).

Nakamura, Y., Inagaki, Y., & Nakahara, T. (2014a). Development of math input interface for STACK by utilizing MathDox. In *Proceedings of 2014 PC Conference* (pp. 188–191) (in Japanese).

Nakamura, Y., Taniguchi, T., & Nakahara, T. (2014b). Item bank system for the mathematics e-learning system STACK. *Electronic Journal of Mathematics & Technology, 8*(5), 355–362.

Nakamura, Y., & Nakahara, T. (2016). Development of a math input interface with flick operation for mobile devices. In *Proceedings of 12th International Conference Mobile Learning 2016* (pp. 113–116).

Nakamura, Y., & Nakahara, T. (2017). A new math input interface with flick operaton for mobile devices. *MSOR Connections, 15*(2), 76–82.

Nakamura, Y., Ohmata, Y., & Nakahara, T. (2012). Development of a question-authoring tool for math e-learning system stack. In *Proceedings of IADIS International Conference E-Learning* (pp. 435–440).

Osaka Prefecture University (2016). *MATH ON WEB: Learning college mathematics by webMathematica.* http://www.las.osakafu-u.ac.jp/lecture/math/MathOnWeb/.

Pollanen, M., Wisniewski, T., & Yu, X. (2007). Xpress: A novice interface for the real-time communication of mathematical expressions. In *Proceedings of MathUI 2007.* Available at http://euclid.trentu.ca/math/marco/papers/3_Pollanen-Xpress.pdf. Accessed July 12, 2017.

Reed, C. (2013). *Creating, visualizing, and exploring knowledge maps.* http://vis.berkeley.edu/courses/cs294-10-fa13/wiki/images/e/eb/Final_poster_colorado_reed.pdf. Accessed July 14, 2017.

Sangwin, C. (2012). The dragmath equation editor. *MSOR Connections.*

Sangwin, C. (2013). *Computer aided assessment of mathematics.* Oxford: Oxford University Press.

Shirai, S., & Fukui, T. (2014). Improvement in the input of mathematical formulae into STACK using interactive methodology. *Computer & Education, 37,* 85–90 (in Japanese).

Shirai, S., & Fukui, T. (2016). *MathTOUCH Web*. http://math.mukogawa-u.ac.jp/en/. Accessed July 12, 2017.

Shirai, S., & Fukui, T. (2017). MathTOUCH: Mathematical input interface for e-assessment systems. *MSOR Connections, 15*(2), 70–75.

Shirai, S., Nakamura, Y., & Fukui, T. (2015). An interactive math input method for computer aided assessment systems in mathematics. *IPSJ Transactions on Computers and Education, 1* (3), 11–21 (in Japanese).

Smithies, S., Novins, K., & Arvo, J. (2001). Equation entry and editing via handwriting and gesture recognition. *Behaviour and Information Technology, 20*(1), 53–67.

Taniguchi, T., Udagawa, S., Nakamura, Y., & Nakahara, T. (2015). Math questions of differential equation, gamma function and beta function using STACK on moodle. *RIMS Kokyuroku, 1978,* 79–86 (in Japanese).

Thaule, M. (2016). Maple T.A. as an integrated part of calculus courses for engineering students. In *Maple T.A. and Möbius User Summit 2016*.

Wood, K. (2014). *jQuery Keypad (2.0.1)*. [Computer software]. Retrieved from http://keith-wood. name/keypad.html. Accessed July 12, 2017.

Yoshitomi, K. (2014). On a formulation of "Mathematics e-Learning Contents Specification" and it's applications to some systems. In *Proceedings of the Annual Conference of JSiSE* (Vol. 39. pp. 167–168) (in Japanese).

Yoshitomi, K., & Kawazoe, M. (2013). On the framework of database for e-learning contents of mathematics. *JSiSE Research Report, 28*(1), 23–28 (in Japanese).

Chapter 9
Designing Interactive Technology to Scaffold Generative Pedagogical Practice

Anthony Matranga, Jason Silverman, Valerie Klein
and Wesley Shumar

Abstract This chapter introduces a web-based assessment environment, the EnCoMPASS Environment, that was purposefully designed to scaffold activities consistent with a group of mathematics teacher educators' practices as well as research-based instructional practices. The chapter details the design of the tool and then presents preliminary findings from our analysis of 21 practicing teachers' collective mathematical activity mediated by the tool. Findings indicate that the software environment supported teachers' participation in common practices for examining student work as well as more generative practices such as providing evidence-based feedback. The study has implications for a way in which to conceive of the design of technologies to support generative professional development at a distance.

Keywords Technologically mediated professional develop · Software design
Teacher professional development

A. Matranga (✉)
California State University San Marcos, San Marcos, USA
e-mail: amatranga@csusm.edu

J. Silverman · V. Klein · W. Shumar
School of Education, Drexel University, 3401 Market Street,
Room 358, Philadelphia, PA 19104, USA
e-mail: silverman@drexel.edu

V. Klein
e-mail: vek25@drexel.edu

W. Shumar
e-mail: shumarw@drexel.edu

© Springer International Publishing AG, part of Springer Nature 2018   149
J. Silverman and V. Hoyos (eds.), *Distance Learning, E-Learning and Blended Learning in Mathematics Education*, ICME-13 Monographs,
https://doi.org/10.1007/978-3-319-90790-1_9

9.1 Introduction

There is a collective effort amongst mathematics education researchers to develop and refine ways in which to support mathematics teachers' instructional change. It is widely accepted that professional development (PD) is an effective approach to impacting teachers' instruction. There are a variety of approaches to PD that have shown potential to support teachers' instructional change such as PD where teachers plan, rehearse and analyze classroom instruction with teacher educators (Lampert et al., 2013), examine records of practice (i.e. videos of classroom interactions (Sherin, 2007) or student mathematical thinking Jacobs, Lamb, & Philipp, 2010), or participate in communities with generative and productive norms (McLaughlin & Talbert, 2001). There is evidence that community-based PD, in particular, is effective in supporting teachers prolonged and generative change (Vescio, Ross, & Adams, 2008). With advances in technology, research is beginning to investigate the potential for teacher professional development in online spaces (Goos & Bennison, 2008; Matranga, 2017; Trust, Krutka, & Carpenter, 2016). Online communities enhance access to high quality professional development and allow teachers to fit community into their daily schedule. Research also indicates that norms that emerge in alternative contexts are transferable into teachers' instructional practice, even if they come in conflict with instructional norms in teachers' local schools and districts (Vescio et al., 2008).

There is little research that focuses on how to support the emergence of online communities and in particular how to support the emergence of communities that engage particular norms and instructional practices. Our current work aims to address this gap in the literature through the design of an enhanced web-based assessment environment that can scaffold teachers' participation in particular activities that are consistent with a community of teacher educators' practices. The broad goal of facilitating teachers' work with the web-based assessment environment is to support the emergence of generative and productive norms that could transfer into teachers' instructional practice and engender a more student-centered learning environment. This chapter introduces the design of this web-based tool and discusses emerging results from a case study in which we analyzed teachers' use of the tool in the context of an online community-based PD course for practicing teachers.

This chapter is organized as follows. First, we discuss our conceptualization of professional development in order to motivate the design of the tool. Second, we discuss the Math Forum—an online community for mathematics and mathematics education—and their core practices. Third, the web-based tool is introduced in which the design features are intended to scaffold activities consistent with the Math Forum's practices. Fourth, we discuss emerging results from our analysis of teachers' use of the tool. The chapter concludes with a discussion regarding the implications this work has for the design of PD and enhancing mathematics teachers' instruction.

9.2 Designing the EnCoMPASS Environment

Our work is grounded in sociocultural theories of learning and in particular communities of practice framework that takes evidence of learning as increasing one's participation in a community of practice (Wenger, 1998). Wenger (1998) argues that social life involves participation in multiple communities of practice, where involvement in a community includes engaging shared practices, having common goals and a shared set of tools. One consequence of this perspective of learning is that as individuals engage in practice with members of a particular community, boundaries often form between those who have been participating in the community and those who have not been participating in the community. Because of this phenomenon, one way to conceptualize engineering learning experiences is through bridging communities, where members of different communities come together and engage in collective practice, thereby engaging a boundary encounter.

Sztajn, Wilson, Edgington, and Myers (2014) conceptualize mathematics teacher professional development as boundary encounters between communities of teachers and communities of teacher educators. Teachers and teacher educators can be conceived of as participating in different communities of practice, as they engage different practices around analyzing and making sense of student thinking. In this sense, Sztajn et al. (2014) argues professional development should be practice-based, where members of these communities are engaging practice around artifacts of teaching.

The concept of boundary objects is used to conceive of artifacts that have potential to support generative work at the boundary between communities. Boundary objects are objects or environments originally conceptualized as effective in mediating activity in the absence of consensus (Star & Griesemer, 1989). One of the properties of boundary objects is interpretive flexibility, that is the potential of an object's perceived use to vary according to the communities in which are engaging with the object (Star, 2010). An artifact with interpretive flexibility has the potential to engender a generative learning environment because when different communities come together and engage practice around the object it is likely that differences in perspective will arise affording opportunities for negotiation and the transformation of practice. Thus, PD activities that include interactions between teachers and teacher educators mediated by a boundary object have the potential to provoke generative conversations.

Our work intended to design a web-based software environment that can function as a boundary object and mediate generative work between communities of teachers and communities of teacher educators. Building on extant research around boundary objects, we conceptualized the design of a tool that could have the same generativity as a boundary object, while situated within a context in which only members of a teacher community are interacting with one another. In this sense, we intended to emulate a boundary encounter between a community of teachers and a community of teacher educators by mediating a group of teachers' work with a software environment that would function as a boundary object but also scaffold participation in activities consistent with the Math Forum's practices. Thus, documenting the Math Forum's core practices was an important part of the design of the web-based tool. The following section introduces the Math Forum and provides an overview of two of the Math Forum's core practices.

9.2.1 The Math Forum

The Math Forum is a website for mathematics and mathematics education as well as a community of mathematics teacher educators. The Math Forum's website houses services and digital archives designed to mediate communication on the Internet about mathematics as well as to provide resources for teachers when planning instruction. The Math Forum staff are a group of teacher educators who travel the US and conduct workshops with mathematics teachers and promote student-centered instruction consistent with instructional practices called for by the NCTM (e.g. orchestrating rich mathematical discussions, scaffolding peer-to-peer argumentation, etc.) (NCTM, 2000).

In our work with the Math Forum over the last two decades we have documented what we refer to as the Math Forum's core practices. One of these core practices we refer to as *valuing*. Valuing is grounded in the belief that "individuals have great things to contribute" (Renninger & Shumar, 2004, p. 197) both mathematically and otherwise. Valuing is operationalized in the Math Forum's activity of noticing and wondering. Noticing and wondering at the Math Forum originated in staff's PD work with teachers as a way to frame the ways in which they looked at student work (Shumar & Klein, 2016). Noticing frames interrogation of students' ideas as a way to attend to the mathematical details of students' thinking and then wondering is a process of grounding analysis in students' thinking by asking specific questions. This activity is at the core of the way in which the Math Forum works to understand students' mathematical thinking.

The second core practice of the Math Forum is providing *evidence-based feedback*. This practice of the Math Forum can be likened to a "research lens," or a process of developing and testing conjectures to better improve conditions for learning. Providing evidence-based feedback includes two activities: (1) collecting evidence of student thinking through processes of noticing and wondering, and then (2) reflecting upon this initial layer of analysis to parse through noticings and wonderings as a means to target aspects of student thinking that are likely inchoate forms of significant mathematical understandings. The Math Forum staff use these activities to prepare to design feedback that can create an environment for students to expand their mathematical understandings. Following providing feedback to students, the Math Forum staff reengage these activities to further understand student mathematical thinking and support learning.

In regard to the broader landscape of mathematics education research, the Math Forum's core practices are consistent with the NCTM's principles and standards for mathematics as well as research-based instructional practices that advocate student-centered instructional strategies. *Valuing* and providing *evidence-based feedback* are consistent with calls by the NCTM to provide **all** students opportunities to engage rigorous mathematical thought (NCTM, 2000). Valuing students' ideas by focusing on the details of their thinking and grounding analysis within this thinking is a way to take *each and every* student's thinking seriously. Moreover, providing evidence-based feedback is a way to support *each* student in expanding his or her current way of knowing through linking feedback to that student's mathematical thinking. In addition, valuing and providing evidence-based feedback are consistent with practices such as professional noticing (Jacobs et al., 2010) and formative assessment (Heritage, Kim, Vendlinski, & Herman, 2009). The professional noticing framework involves attending to the details of student thinking and interpreting the meaning of these details for students' mathematical understandings while developing this understanding of student thinking is a starting point for designing feedback and learning environments that support student mathematics learning.

Taken together, the Math Forum's practices are at the core of their success as a community of mathematics teacher educators and their practices are consistent with those called for by research and policy to improve mathematics education in the United States. Thus, our work began with the conjecture that designing a web-based assessment environment that could scaffold activities consistent with the Math Forum's practices has the potential to improve teachers' mathematics instruction to become more consistent with what is called for by research and policy.

9.2.2   The EnCoMPASS Environment

The EnCoMPASS Environment is designed to function as a boundary object through affording participation in teachers' existing practices for organizing and assessing student work and it is also designed to scaffold activities for examining student work consistent with the Math Forum's practice of valuing and providing evidence-based feedback. Thus, the following introduces (1) the landscape of the EnCoMPASS Environment and its design features, (2) the way in which these features scaffold activities consistent with the Math Forum's practices, and (3) the way in which the tool is designed to function as a boundary object by affording participation in teachers' existing ways of examining student work.

9.2.2.1   The Landscape of the EnCoMPASS Environment

The EnCoMPASS Environment is a web-based assessment environment that provides a space for teachers to upload sets of student work into a primary workspace (shown in Fig. 9.1). The design features of the EnCoMPASS Environment are intended to enhance the process of looking at student work and developing feedback. The primary workspace is separated into three panels. Student work is populated into the middle panel while the left and right panels scaffold the analysis of students' work. The features that support this analysis include a selection tool, a noticing and wondering commenting tool (center pane), the capacity to sort and categorize (left pane) and aggregate for future use and feedback (right pane).

The *selection tool* is designed to scaffold a process of highlighting aspects of students' work. Selections are collected at the bottom of the screen below the student's work (shown in yellow in Fig. 9.1). In the right panel, there is a text field that provides space for teachers to record their thinking. These comments are in the form of "*I notice...*" and "*I wonder...*" and are directly linked to selected aspects of the student's work. Below the text field is a list of the noticings and wonderings, which are available for reuse. In the left panel, there is a *categorization system* or folders, which facilitate organization of selections and comments. This feature allows teachers to develop a set of folders and sort selections based on different characteristics (i.e. strategy used to solve the problem, completeness, correctness, etc.). Lastly, the EnCoMPASS Environment has an *aggregation system* that organizes teachers' selections, noticings and wonderings for a specific students' work and organizes them into the *feedback screen* (shown in Fig. 9.2). For example, in Fig. 9.2, following "you wrote:" is an aspect of student work that the teacher highlighted using the selection tool. Moreover, following "...and I noticed that..." is the comment the teacher made using the noticing and wondering commenting field on that particular selection. In this screen, teachers can also edit their selections/noticings/wonderings to develop a coherent note that is sent to a student.

While this section introduced the functionality of the EnCoMPASS Environment's features, the following relates the activities in which these features scaffold to the Math Forum's practices.

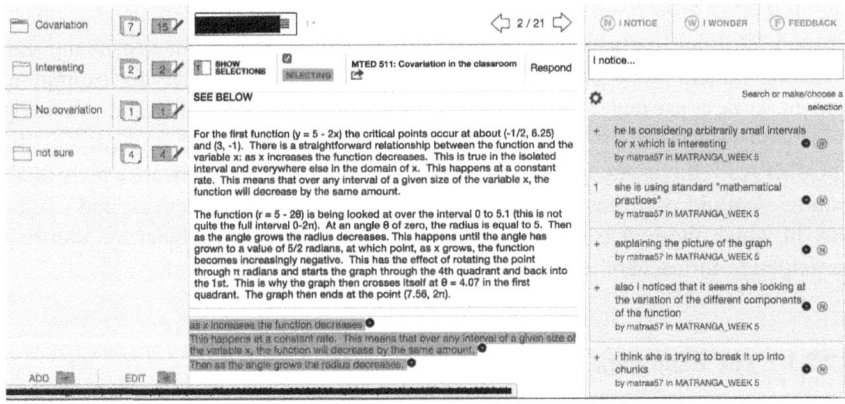

Fig. 9.1 Primary workspace in the EnCoMPASS environment (color figure online)

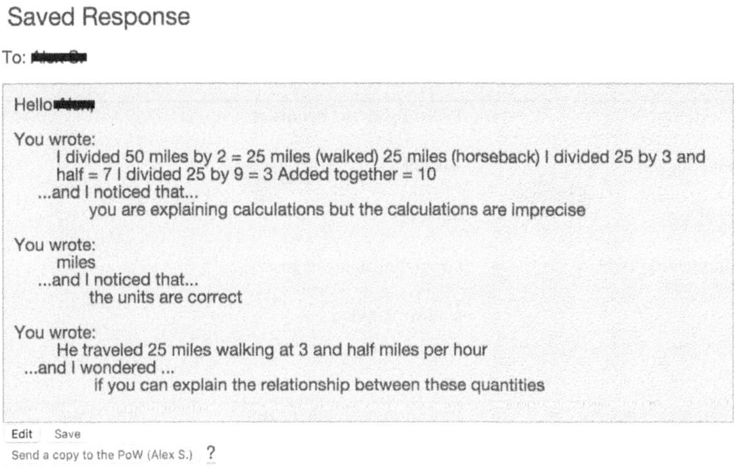

Saved Response

To: ▰▰▰▰

> Hello▰▰▰▰
>
> You wrote:
>   I divided 50 miles by 2 = 25 miles (walked) 25 miles (horseback) I divided 25 by 3 and
>   half = 7 I divided 25 by 9 = 3 Added together = 10
> ...and I noticed that...
>   you are explaining calculations but the calculations are imprecise
>
> You wrote:
>   miles
> ...and I noticed that...
>   the units are correct
>
> You wrote:
>   He traveled 25 miles walking at 3 and half miles per hour
> ...and I wondered ...
>   if you can explain the relationship between these quantities

Edit   Save
Send a copy to the PoW (Alex S.)   ?

Fig. 9.2   The feedback screen

9.2.2.2   Scaffolding Activities Consistent with the Math Forum's Practices

Table 9.1 provides an overview of the conjectures that guided the design of the EnCoMPASS Environment and this section details how the features of the environment scaffold activities consistent with the Math Forum's practices of valuing and providing evidence-based feedback. The Math Forum's practice of *valuing* takes seriously the notion that everyone has something to contribute to a conversation and is operationalized through the activity of noticing and wondering. Noticing and wondering includes activities such as focusing on the details of student thinking and grounding analysis in these details by asking specific questions. The *noticing and wondering commenting field* and *selection tool* are designed to scaffold the activities just mentioned.

The noticing and wondering commenting tool provides an entry point into focusing on the details of student work and grounding analysis in these details. Noticing frames analysis of student work through the lens of "I notice," which is intended to focus user's attention on anything that is interesting, unique or questionable. Once an aspect of student work is "noticed," framing additional thinking through "I wonder" is intended to scaffold careful thinking about what the "notice" or evidence of student thinking could say about the students' mathematical understandings. In this sense, the activity of noticing and wondering sets a frame around which the user engages with the selection tool.

The primary activity in which the selection tool supports as well as its underlying functionality further scaffold activities of focusing on the details of student thinking and grounding analysis within this thinking. In particular, the selection tool scaffolds these activities by supporting the process of "selecting" or "highlighting" aspects of students' work noticed by the analyst. As aspects of student thinking are

Table 9.1 Conjectured relationship between features, activity and practice

Design features	Activity	Math Forum practice
• Selection tool • Noticing and wondering commenting field	• Focusing on the details of student thinking • Grounding analysis in student thinking	Valuing
• Aggregation system	• Reflect upon evidence of student thinking to develop focused feedback	Evidence-based feedback

selected, they are aggregated at the bottom of the screen (see Fig. 9.1—shown in yellow at the bottom of the center panel in the primary workspace). This isolates instances of student thinking and affords the opportunity for additional thinking to be done by the analyst about these details as well for the analyst's thoughts to be recorded with the noticing and wondering commenting field. In addition, the EnCoMPASS Environment generates a link between the selection and comment, thus scaffolding the grounding of analysis in student thinking. In particular, in order to record a notice or wonder in the text field one of the highlighted aspects of student work collected at the bottom of the screen must be 'clicked' prior to recording the comment in the noticing and wondering commenting field. Moreover, once a comment is made, if the user clicks on a comment from the list of comments in the right panel of the primary workspace the corresponding selection highlighted in yellow at the bottom of the center panel in which this comment was connected is underlined in red. To this end, the EnCoMPASS Environment is designed to mediate activities involved in noticing and wondering and, consequently, *valuing* by focusing user's analysis on details of student thinking as well as by grounding this analysis in student thinking. While these aspects of the tool's design are intended to support the analysis of student work, the aggregation system is intended to support the design of feedback in ways that are consistent with the Math Forum's practice of providing evidence-based feedback.

In fact, the Math Forum's practice of providing *evidence-based feedback* is regarded as a 'research lens' for examining student thinking, which includes two activities (1) collecting evidence of student thinking through process of noticing and wondering, and (2) reflecting upon this initial layer of analysis to develop focused feedback.

The selection tool and noticing and wondering commenting field of the EnCoMPASS Environment are designed to scaffold activity consistent with the initial process of providing evidence-based feedback. Thus, the initial layer of analysis of student work results in a collection of highlighted aspects of student thinking and noticings/wonderings that are explicitly linked to this data. The EnCoMPASS Environment therefore creates residue of this first pass of analysis of student thinking and then when the user is ready to develop feedback, the *aggregation system* transitions users to the *feedback screen* (shown in Fig. 9.2), which aggregates selections/noticings/wonderings in order to provide a snapshot of the thinking done in the initial analysis. This screen scaffolds reflection on the initial layer of analysis as there is an "edit button" that allows users to adjust, reword, reorganize and build upon the thinking done during the initial analysis. In this way, the feedback is grounded in student thinking as users are supported in transforming evidence of student thinking and documentation of their own thinking that is linked to this evidence into a focused feedback note designed to support students in expanding their mathematical ways of knowing.

9.2.2.3 Teachers' Existing Practice

In addition to scaffolding activity consistent with the Math Forum's practices, the EnCoMPASS Environment was also designed to afford participation in teachers' common practices for preparing for and providing students feedback on their mathematics work. With decades of experience working with teachers, we have found that when presented with a pile of student work teachers (1) sort the work into different piles, (2) assess students' mathematics work, and (3) provide feedback to students based on previous experiences. Sorting student work includes placing students' papers into piles according to particular commonalities in their work. For example, pile A might be 'correct,' pile B might be 'incorrect' and so on. The categorization system of the EnCoMPASS Environment affords participation in this practice as teachers can quickly scan through student work and then place it into folders that are named according to the particular commonality in their work. We also have found that teachers tend to assess student work by circling aspects of a student's work and making brief comments about the particular mistake. The selection tool and noticing and wondering commenting field affords participation in these activities as teachers could highlight, for instance, a calculation error and then comment about what went wrong or how to fix the error. The tool also affords the development of feedback according to teachers' experiences as they could look at student work and then transition directly to the feedback screen without using the selection tool or the noticing and wondering commenting field.

The way in which the EnCoMPASS Environment affords participation in practices for preparing for and providing students feedback on their mathematics work is consistent with what we have found is typical for practicing teachers and is important for the tool to function as a boundary object. If the tool was not designed to have interpretive flexibility and support participation in such activities, it is unlikely that teachers would perceive the tool as useful and might not legitimately engage with the EnCoMPASS Environment.

9.3 Examining Teachers' Interactions Mediated by the EnCoMPASS Environment

In one of our initial use cases of the EnCoMPASS Environment, we integrated the tool into an online community-based PD course for practicing teachers that included exclusively asynchronous communication. The existing structure for engaging collaborative problem solving in the course included providing teachers with a problem in which they would spend 3–4 days to work privately on drafting a response and then post their response to the course. Teachers would then review their colleagues' work and provide them feedback. The final stage included revising the initial submission according to their colleagues' feedback. Modifying this process for this study, participants uploaded their colleagues' work into the EnCoMPASS Environment and then used the web-based tool to scaffold the process of providing their colleagues feedback.

The current study included 21 practicing teachers who participated in the online PD course. The participants were primarily novice teachers that ranged from only having student teaching experiences to three years of experience in the classroom. The analysis in this study used a grounded theory methodology (Glaser & Strauss, 1999) where we conducted open and axial coding procedures with participants' mathematics work and the EnCoMPASS-scaffolded feedback they developed.

9.3.1 Findings

Participants used the EnCoMPASS Environment to examine their colleagues' work and provide feedback for seven of the ten weeks of the course. The tool was introduced in week three and then participants used the tool for each of the following weeks except week seven and ten. Week seven and ten did not include problem-solving activities in which participants used the EnCoMPASS Environment to provide one another feedback because in week seven a group assignment replaced the typical mathematical activities and week ten was reserved for reflective activities. During each of the weeks in which participants used the EnCoMPASS Environment, they examined two of their colleagues' work via the tool and then sent the result of this analysis to their colleague as feedback. As we examined the ways in which participants engaged this process scaffolded by the tool we began to recognize several patterns. We identified a pattern in the connection between the aspects of the mathematics work in which participants were highlighting and the comments they made about these highlights as well as a pattern in the feedback participants were crafting using the noticing and wondering commenting tool. The following briefly examines these patterns, however prior to doing so, we first show that participants used to the tool for its designed use.

9.3.1.1 Using the Tool's Features for Their Intended Use

Through the analysis of participants' interactions mediated by the EnCoMPASS Environment's design features, there is evidence that the tool scaffolded participants' activity in the activities in which it was intended to scaffold. Participants used the selection tool to highlight details of their colleagues' mathematics work and then used the noticing and wondering commenting tool to develop comments connected to these details. Following the use of these features, participants sent the result of their analysis to their colleagues as feedback. An example of the feedback participants developed is shown in Fig. 9.3.

The reader will notice in Fig. 9.3 that the feedback is in the form of a list of text labeled "You wrote…," "…and I noticed that…" or "…and I wondered about…" The EnCoMPASS Environment generates these labels. "You wrote:" signifies the particular instance of mathematical thinking a participant highlighted using the selection tool while "…and I noticed/wondered that/about…" signify the comment participants made using the noticing and wondering commenting tool. Given the "look" of participants' feedback (in which a representative example is shown in

```
You wrote: relationship between a length x and the area of a rectangle with sides 2x and
    3x
    ... and I noticed that ... these are the quantities you are focusing on
You wrote: area of a rectangle with sides 2x and 3 is dependent on the value of the
    quantity x.
    ... and I noticed that ... quantities you are working with and the relationship
You wrote: x-intercept and a y-intercept of zero. This is a result of the fact that if the
    length X is zero than the area must be zero.
    ... and I noticed that ... I made the connection of the area being zero when the
        length of x was 0, but I didn't relate that to the x and y intercept. Nice
        connection
You wrote: the vertex is a minimum because the area cannot be negative therefore the
    smallest area possible is zero.
    ... and I noticed that ... good explanation of this value
You wrote: parabola with an axis of symmetry through x=0
    ... and I wondered about ... if you could explain the axis of symmetry. It may help in
        student understanding.
You wrote: As x is squared, this allows for the graph to exist when x is negative even
    though we do not measure negative lengths.
    ... and I noticed that ... This is a great way to explain the negative length. I used
        the idea that we were measuring in the opposite direction so the negative
        represents direction.
```

Fig. 9.3 Example feedback developed scaffolded by the EnCoMPASS environment

Fig. 9.3), it appears that the tool scaffolded participant's development of feedback in the way in which it was designed to because participants selected the details of their colleagues' work and then developed comments connected to these details.

To place this use case in contrast to one that would have come in conflict with the EnCoMPASS Environment's intended use, participants could have used the selection tool to select the entirety of their colleague's initial response and then make broad comments about their colleague's work using the noticing and wondering commenting tool. Moreover, participants could have used the selection tool to highlight details of their colleague's work and then transitioned directly to the feedback screen to provide feedback without using the noticing and wondering commenting tool. Nevertheless, the above example shown in Fig. 9.3 provides evidence that participants used the EnCoMPASS Environment for its designed use.

9.3.2 Linking Comments to Data

Closer analysis of participants' feedback indicates that participants explicitly linked their noticing and wonderings to evidence of their colleagues' thinking and they did so in two ways. First, participants used pronouns such as "this" or "here" to refer to the highlighted portion of their colleague's work in which they were referring to in the comment they developed with the noticing and wondering commenting tool. Moreover, participants linked the content of the highlighted aspect of their colleague's work with the content of their noticing/wondering. For example, consider the following representative example from an occasion where the class is making sense of the quantities from the unit circle (e.g. arc length, vertical/horizontal distances from the circumference of the circle to the axes) in order to make sense of the behavior of the sine function.

(Jazmine's selection from Rose's work) You wrote: domain

(Jazmine's Comment on the above selection)…and I noticed that… you used the word domain; I don't think I did

(Jazmine's selection from Rose's work) You wrote: x represents the angle or the arc length of the circle

(Jazmine's Comment on the above selection)…and I wondered about… I think this is different from my explanation. I wonder if one of us is correct; or if we are both correct, but saying it differently.

In the first selection, Jazmine highlighted Rose's use of the word 'domain' from her solution and then in Jazmine's comment on this selection she noticed that Rose used the word domain. Jazmine's second selection highlighted Rose's description of a particular quantity ("the angle or arc length…") and then in her comment on this selection, Jazmine wondered, "I think *this* [emphasis added] is different…".

This example illustrates the way in which Jazmine linked her comments to the selections she made from Rose's work. First, Jazmine used the design features for their intended use as she selected the details of Rose's work. In Jazmine's first comment, there was a link between the content of Rose's work and the content of her comment, namely the word "domain." In Jazmine's second comment, she explicitly referenced Rose's work with use of the word "this." Thus, in both cases there was a specific highlighted detail of Rose's work and an explicit link between this detail and the comment. In particular the link was through (1) common use of terms (e.g. domain) and (2) pronoun usage to refer to the selection.

This was the typical way in which participants provided feedback to colleagues scaffolded by the EnCoMPASS Environment, which was consistent with the operationalization of *valuing* through noticing and wondering. As noted above, noticing and wondering includes focusing on the details of student work and then grounding analysis in these details. Thus, there is evidence that the EnCoMPASS Environment scaffolded activities that are consistent with the Math Forum's practice of *valuing*, as participants highlighted the details of colleagues' work and then linked their comments to this evidence of their colleagues' thinking.

9.3.2.1   An Emerging Purpose for Feedback

Emerging from participants' use of the selection tool and noticing and wondering commenting field was a pattern in the feedback they provided to colleagues that had the purpose of challenging colleagues to refine the details of their mathematical explanations. When challenging colleagues, participants linked their feedback to data and explicitly asked colleagues to further refine and expand upon their mathematical explanations. To illustrate this use of the tool, an example is taken from an activity where participants were working with the function $y = \sin(x)$. Consistent with the goals of the course, participants were attempting to examine the relationship between quantities to make sense of the behavior of $y = \sin(x)$. There were a number of cases where participants develop explanations that were not consistent with the goals of the course, which invoked occasions where participants would challenge colleagues who developed such explanations. The following illustrates how Paul used the tool to develop feedback that challenges Nina to refine her mathematical explanation.

> **Paul's selection from Nina's work:** You wrote: This graph appears as it does because of the Unit Circle. Essentially as the values of $\sin(x)$ make their way around the circle, they start again at zero.
>
> **Paul's comment to Nina:**...and I wonder... if you could elaborate on this concept more. Why do the values start again at zero? Why does the graph have hills and valleys?

Using the selection tool of the EnCoMPASS Environment, Paul highlighted an aspect of Nina's work and then made a comment grounded in this detail as he used "this" to refer to Nina's work when Paul said, "if you could elaborate on *this* concept more." In his comment, Paul challenged Nina to expand her mathematical explanation when he said, "Why do the values start again at zero?" "Why does the graph have hills and valleys?" Part of the reason why this was regarded as a challenge is because the class was working collectively to explain why graphs look a particular way and it appears that Nina did not include such a description in her explanation.

While the EnCoMPASS Environment scaffolded participants' examination of the details of colleagues' thinking and grounding comments within those details, there was nothing inherent about the tool's design features that scaffolded challenging colleagues. Therefore, it appears that challenging was emergent, in that the purpose of the feedback emerged through the use of the EnCoMPASS Environment for developing feedback.

Taken together, the brief examination of participants' use of the EnCoMPASS Environment for developing feedback illustrates that (1) participants used the design features for their intended use as they made selections and made comments connected to these selections, (2) the tool scaffolded activities consistent with the Math Forum's practice of *valuing* as participants began to explicitly link their noticings and wonderings to data, and (3) the purpose of participants' feedback was emergent in that the tool was not designed to scaffold challenging colleagues to refine their mathematical explanations.

9.4 Discussion

The intention of the design of the EnCoMPASS Environment is to scaffold generative and productive norms for preparing for and providing students feedback on their mathematics work (consistent with the Math Forum's practices) that can transfer into teachers' instructional practice and engender a more student-centered learning environment. This study found that the design features of the EnCoMPASS Environment scaffolded activities in which they were designed to scaffold. Moreover, as a result of this activity, participants began to engage practices for preparing and providing feedback to students in ways in which were consistent with the Math Forum's practice of valuing and developing generative feedback through challenging colleagues. This result suggests the potential of the EnCoMPASS Environment to scaffold generative work between teachers in online community-based PD.

Earlier in this chapter, we mentioned that research indicates that norms that emerge in alternative contexts are transferable into teachers' instruction. While this study did not document the emergence of norms, as a result of participation in this study it is more likely that participants would focus on the details of their students' thinking and then link feedback to this data. Moreover, there was likely an increased potential for participants to challenge students to refine their mathematical explanations. In this sense, students' ideas would become more central to teachers' instruction as teachers use student thinking as the foundation on which they think about how to respond to students and move the class forward in their thinking. Thus, there is potential that participation in community-based PD mediated by the EnCoMPASS Environment can support teachers in moving along a trajectory from teacher-centered to student-centered instruction.

9.5 Conclusion

This study found that a technology mediating interactions in a collaborative environment had potential to impact teachers' mathematics instruction rather than a group of teacher educators facilitating PD activities. A digital platform, namely the EnCoMPASS Environment, was designed to emulate teacher or student participation within the Math Forum and appeared to have potential impact on the ways in which teachers provide one another feedback on their mathematics work in similar ways in which participation in PD with the Math Forum staff would impact these practices. This suggests that this environment has the potential to impact the norms and practices of an online community of teachers and ostensibly impact teachers' classroom practice. Given that the tool could be integrated into multiple contexts simultaneously, it has the potential to enhance the scale at which the Math Forum could impact mathematics teachers' instruction.

While there is emerging evidence that this tool began to scaffold participation in generative and productive norms for providing feedback, we are still in the process of analyzing data to make sense of how the tool's design to function as a boundary object impacted teachers' use of the EnCoMPASS Environment. At this phase of the analysis, we have preliminary conjectures that emerged through observations from facilitating teachers' use of the tool in the online PD course. In particular, we observed teachers expressing affect towards the design of the tool and its potential to make analysis of student work more efficient.

In summary, at this stage of our work, there is evidence that the EnCoMPASS Environment is functioning as a boundary object because it is perceived as legitimate and the design features have interpretive flexibility through their use to share and compare information as well as to challenge colleagues. Given our findings from the study presented in this chapter, we argue that conceptualizing the design of technology as a boundary object is one way in which to conceive of a scalable design for collaborative and technologically mediated professional development that takes place at a distance. Future research is needed to better understand how the EnCoMPASS Environment functions as a boundary object and how functioning as such is significant for scaffolding participation in activities consistent with the Math Forum's practices.

References

Glaser, B., & Strauss, A. L. (1999). *The discovery of grounded theory: Strategies for qualitative researcher*. Hawthorne, NY: Aldine de Gruyter.

Goos, M. E., & Bennison, A. (2008). Developing a communal identity as beginning teachers of mathematics: Emergence of an online community of practice. *Journal of Mathematics Teacher Education, 11*(1), 41–60.

Heritage, M., Kim, J., Vendlinski, T., & Herman, J. (2009). From evidence to action: A seamless process in formative assessment? *Educational Measurement: Issues and Practice, 28*(3), 24–31.

Jacobs, V. R., Lamb, L. L., & Philipp, R. A. (2010). Professional noticing of children's mathematical thinking. *Journal for Research in Mathematics Education, 41,* 169–202.

Lampert, M., Franke, M. L., Kazemi, E., Ghousseini, H., Turrou, A. C., Beasley, H., ... Crowe, K. (2013). Keeping it complex: Using rehearsals to support novice teacher learning of ambitious teaching. *Journal of Teacher Education, 64*(3), 226–243.

Matranga, A. (2017). *Mathematics teacher professional development as a virtual boundary encounter* (Unpublished doctoral dissertation). School of Education, Drexel University, Philadelphia, PA.

McLaughlin, M. W., & Talbert, J. E. (2001). *Professional communities and the work of high school teaching.* University of Chicago Press.

NCTM. (2000). *Exucitive summary: Principle and standards for school mathematics.* Reston, VA: National Council of Teachers of Mathematics.

Renninger, K. A., & Shumar, W. (2004). The centrality of culture and community to participant learning at and with The Math Forum. In: S. Barab, R. Kling, & J. H. Gray (Eds.), *Designing for virtual communities in the service of learning* (pp. 181–209). Cambridge, MA: Cambridge University Press.

Sherin, M. G. (2007). The development of teachers' professional vision in video clubs. In *Video research in the learning sciences* (pp. 383–395). Hillsdale, NJ: Erlbaum.

Shumar, W., & Klein, V. (2016). *Technologically mediated noticing & wondering @ the math forum.* Paper presented at the Annual Meeting of the American Educational Research Association, Washington, DC.

Star, S. L. (2010). This is not a boundary object: Reflections on the origin of a concept. *Science, Technology and Human Values, 35*(5), 601–617.

Star, S. L., & Griesemer, J. R. (1989). Institutional ecology, translations' and boundary objects: Amateurs and professionals in Berkeley's Museum of Vertebrate Zoology, 1907–39. *Social Studies of Science, 19*(3), 387–420.

Sztajn, P., Wilson, P. H., Edgington, C., & Myers, M. (2014). Mathematics professional development as design for boundary encounters. *ZDM Mathematics Education, 46*(2), 201–212.

Trust, T., Krutka, D. G., & Carpenter, J. P. (2016). "Together we are better": Professional learning networks for teachers. *Computers & Education, 102,* 15–34.

Vescio, V., Ross, D., & Adams, A. (2008). A review of research on the impact of professional learning communities on teaching practice and student learning. *Teaching and Teacher Education, 24*(1), 80–91.

Wenger, E. (1998). *Communities of practice: Learning, meaning, and identity.* Cambridge University Press.

Chapter 10
Supporting Teachers in Developing Their RiTPACK Through Using Video Cases in an Online Course

Cosette Crisan

Abstract In order to help the participants on our online course engage critically with research to reflect on whether and how digital technology supports students' understanding and learning of mathematics, a trial focused on the use of online video cases was implemented in a newly designed online course. In this chapter, we report on the use of video cases with the course participants and on the potential of using these videos with the aim of supporting the development of the participants' Research informed Technological Pedagogical Content Knowledge (RiTPACK), with a particular focus on how the digital environment supports students' mathematical work.

Keywords TPACK · Teachers as researchers · Online teaching
Digital technologies

10.1 Introduction

In this chapter I report on a pedagogical intervention in our recently re-developed Masters level online course 'Digital Technologies for Mathematical Learning,' which focuses on the teaching and learning of mathematics supported by digital technologies. The intervention is aimed at supporting the enrolled teachers engage with research as they develop their Technological, Pedagogical and Content Knowledge (TPACK).

C. Crisan (✉)
UCL Institute of Education, University College London,
20 Bedford Way, London WC1H 0AL, UK
e-mail: c.crisan@ucl.ac.uk

© Springer International Publishing AG, part of Springer Nature 2018   165
J. Silverman and V. Hoyos (eds.), *Distance Learning, E-Learning and Blended Learning in Mathematics Education*, ICME-13 Monographs,
https://doi.org/10.1007/978-3-319-90790-1_10

10.1.1 Overview of the Chapter

I start this chapter with a description of the content and organization of the online course. I then discuss the pedagogical principles underlying the design of our online course, describing the rationale for our pedagogical innovation and intervention, namely the use of online video cases. The latter part of this chapter introduces a theoretical framework adapted from the literature in order to account for the course participants' learning as they started experimenting with using the new technology in their teaching practices and linking it with the theoretical and research knowledge base of the course. Finally, a case study is presented, together with the methods employed in the collection and analysis of the data. A discussion of the learning and engagement with research of one of the course participants[1] and a brief conclusion end this chapter.

10.2 Context and Course Description

There are two e-learning aspects of this Masters level course: (1) its online delivery and (2) the e-focus of the course itself, consisting of (i) familiarization of the participants (practicing or prospective mathematics teachers) with a wide range of digital tools and resources (graph plotters, dynamic geometry environments, statistical software, fully interactive online packages) and (ii) critical reflection on the implications of using such tools in the learning and teaching of mathematics at secondary school level (11–18 years old students).

The main aim of this course is to encourage participants to reflect critically on the potential and limitations of digital technologies for the learning and teaching of mathematics by providing opportunities for participants to apply knowledge of relevant research and theory to their professional contexts.

10.2.1 Course Curriculum and Organization

The course is taught online, with participants being given a series of tasks over a ten-week period. The curriculum for this course is divided into three themed sections: Visualizing, Generalizing and Expressing, and Modelling, with each theme lasting for three weeks.

In each of the themed sections, the course curriculum is arranged into a series of short tasks that culminate in the main task of designing and testing a learning activity relevant to each theme. These short weekly tasks are signposted on the

---

[1]In this chapter, I will be referring to the teachers enrolled on our online course as *participants*, while *students* will be used to refer to students in schools.

virtual learning environment of the course (Moodle) at the beginning of each week and include *offline tasks* such as: familiarization with a piece of software and example problems using the specific software; designing a mathematical activity using the specific digital environment; trialling the activity with students or other learners, followed by reflection on the learning episodes. There are also *online tasks* such as: engaging with the ideas in the key readings; reading one of the essential reading articles and writing a response about the points agreed or disagreed with from the article; contribution to online discussion forums, including written observations on views and perspectives of fellow participants. Each theme ends in an activity week, where participants are required to: choose a software tool relevant to the theme, design a learning activity using features of good practice identified from the literature, use the activity they designed with students and analyze its implementation through engagement with research and the ideas assimilated from the literature reviewed to evaluate and justify the implications of using digital technology for students' learning. In each theme, at least one task will form the basis of an online group discussion. The tutors also contribute to these discussions, with the aim of encouraging informed reflection and raising critical awareness of and supporting engagement with the research literature.

10.2.2 Theoretical Background: Pedagogical Underpinnings of Our Online Course

The design of this course has been influenced by the Technological Pedagogical Content Knowledge (TPACK) framework (Mishra & Koehler, 2006) which attempts to describe the body of knowledge and skills needed by a teacher for effective pedagogical practice in a technology enhanced learning and teaching environment. Mishra and Koehler (2006) proposed that a teacher's professional knowledge base *for teaching with the new technology* should include a type of flexible knowledge needed to successfully integrate technology into teaching, informed by and borne out of the interaction of three essential bodies of knowledge: content, pedagogy and technology. Drawing on the work of Koehler and Mishra (2009), Mishra and Koehler (2006), Otrel-Cass, Khoo, and Cowie (2012) described the TPACK as the intersectional relationship of six components as follows (see Table 10.1).

TPACK has been used by many researchers, as this frame offers a helpful way to conceptualize what knowledge teachers need in order to integrate technology into their teaching practice, leaving the specifics of what lies in each circle to disciplinary researchers.

The participants on our course, either practicing or prospective mathematics teachers, bring with them a well-developed or developing PACK (pedagogical and content knowledge base). When designing our course, I planned for opportunities for the participants to familiarize themselves with key types of digital technologies

Table 10.1 TPACK components

TPACK components	Component descriptors
TK or technological knowledge	Understanding about any kind of technological tool
CK or content knowledge	What is known about a given subject
PK or pedagogical knowledge	Teaching methods and processes
PCK or pedagogical content knowledge	Pedagogy specific to a particular subject area or content
TCK or technological content knowledge	What is known about a technology's affordance to represent or enhance content
TPK or technological pedagogical knowledge	Understanding of how technology may support particular teaching approaches

for learning mathematics, at the same time learning to appreciate the rationales and pedagogic strategies associated with these digital technologies for learning mathematics, thus facilitating the development of their TPACK.

10.2.3   Course Evaluation and Reflections

While the participating teachers enrolled in the first presentation of this online course reported development of their TPACK knowledge (as exemplified later in this chapter through a case study), writing about such experiences (as part of their contributions to online forum discussions, as well as part of written tasks and final assignment for this course) and applying the ideas from the key readings in classrooms was a challenge.

Research acknowledges that teachers characterized as 'novice' (with respect to new practices) 'see' less of the complexity of classroom events than experienced teachers do (Yadav & Koehler, 2007). I also realized that the participants in our courses often failed to make connections between their 'research-based' learning with the particular instances of digital technology use in their practices which they were reporting.

I noticed, for example, that during the weekly online discussions, the participants provided narratives of their own learning or classroom based experiences with the new technology. These entries did indeed generate activity on the online forum discussions, but the narratives were mainly about 'what happened'. While this background knowledge was needed in order to comprehend what the learning episode was about, the written format of these asynchronously shared experiences proved to be mainly descriptive, hence time consuming, meaning that the participants rarely reached as far as engaging themselves explicitly with the research and analyze 'why that happened', i.e. how their students' mathematical work was affected by the use of the new technology.

Similarly, for their end of course assignment, the participants were expected to describe, analyze and interpret students' experiences of doing mathematics with digital technology. The analysis of their written assignments provided us with a clear evidence that the participants found it challenging to move beyond description of the learning episodes.

As a tutor, I came to realize that what was needed were opportunities for the participants to engage with analyzing and describing of learning early on in the delivery of the course. I also came to realize that the participants would benefit from shared learning episodes, as this would remove the need for a detailed description of 'what happened', instead allowing participants and tutors to focus on the analysis and interpretation of the learning of mathematics when technology was being used. These reflections led to consideration of a pedagogical intervention in the next presentation of the course, namely providing participants with 'video cases' of students' doing mathematics with digital technology, with the potential to act as a catalyst in generating discussions and reflections focused on the analysis and interpretation of the learning taking place.

10.3 The Study

For the new presentation of the course (starting in January 2016), my intention was also to address Leat, Lofthouse, and Reid (2014) call for the need to develop 'teachers as researchers'. They acknowledge that (worldwide) the relationship teachers have with research is passive, that teachers may or may not choose to use it in their practice. Through the pedagogical intervention mentioned above, my intention was to support the participants in making their conversations more grounded in actual events, more insightful, and more resistant to oversimplifications, thus scaffolding our participants' learning towards more active engagement in undertaking enquiry themselves, which ultimately will benefit their students.

I thus adapted Mishra and Koehler's (2006) TPACK frame to account for the participants' learning as they started experimenting with using the new technology in their teaching practices and linking it with the theoretical and research knowledge base of the course (Fig. 10.1). I refer to this frame as teachers' Research informed Technological Pedagogical Content Knowledge (RiTPACK—my own acronym for this frame).

Learning from my reflection on the first presentation of the course, I decided to provide the participants with *shared episodes* of students' doing mathematics with digital technology and support them in critically analysing and interpreting these episodes by engaging with and making connections with the theory and research they were reading.

Guided by Van Es and Sherin's (2002) study, I considered the use of video cases to provide the participants with a shared learning episode to analyse. Video cases have been used by several mathematics educators and researchers in order to help teachers focus on students' learning and on teachers' decisions made in lessons.

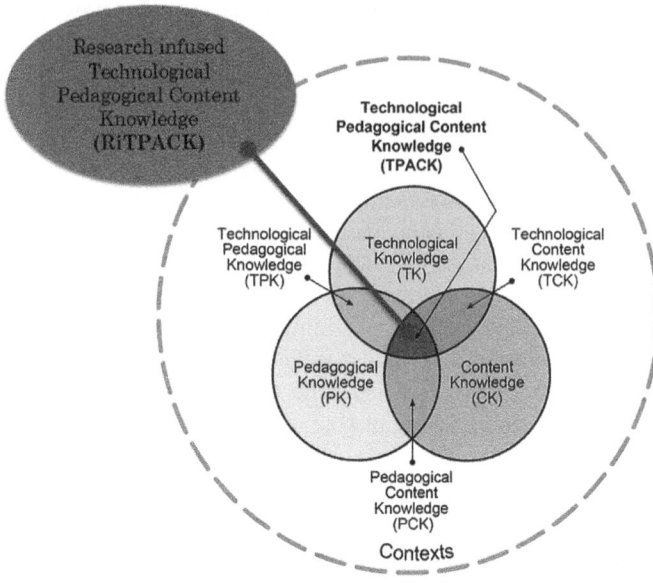

Fig. 10.1 RiTPACK frame

Van Es and Sherin (2002) proposed that videos could be effective tools in helping teachers develop their ability to notice and interpret classroom interactions. Van den Berg (2001) highlights another potential of using videos, namely that they enable teacher educators to prompt the students to watch for specific elements when viewing a video, thus compelling the teachers to look more deeply than they might otherwise have done.

10.3.1 The Aims of the Study

Thus, the aims of this study were to pilot the use of video cases and investigate whether and how this intervention supports and contributes to the development of the participants' RiTPACK, with a particular focus on how the digital environment supports students' mathematical work. My hypotheses was that through such an intervention the participants on the course will be supported in the development of their skills of noticing significant episodes when observing students doing mathematics with the new technology, which they would then analyse and interpret by engaging with the theory and research, with a long term view of preparing them to make informed decisions about use of digital technology that will benefit their students' learning.

10.3.2 Using Video Cases—A Brief Review of Literature

A search through websites which provide video and support materials for those who work in education in the United Kingdom, including teachers, teacher trainers, student teachers and support staff, failed to identify resources with a focus on using digital tools in mathematics lessons. For this reason, in order to support the participants' development of TPACK through reflecting on how the digital environment could support participants' mathematical work, I created and used online video cases in the new presentation of the course. I planned for and recorded a number of videos featuring students working through mathematics activities in a digital environment, referred to as video cases in this chapter.

Of the many features of videos well documented in literature (Calandra, Brantley-Dias, & Lee, 2009; Van Es & Sherin, 2002), one significant feature is the capacity of a video to be paused, rewound, replayed many times in order for the viewer to focus specifically on segments of the videos selected strategically for their significance to the viewer, based on a particular goal (e.g. how the students' learning benefitted (or not) from doing mathematics in a digital environment). The design of the video case was informed by suggestions made by researchers (Van Es & Sherin, 2002), namely that the use of video clips could assist users to shift their attention away from the teachers, the classroom events and evaluating the teaching and learning, and focus it instead onto students' work. Through using a video, teachers can be supported to make tacit ideas explicit because "the process of making images encourages participants to consider why it is that the moment captured on film is important to them" (Liebenberg, 2009, p. 441).

In this research study, the video cases produced are recordings of the work of a pair of students, narrowing the focus of observation on the particular pedagogical activity of noticing significant episodes and analyzing students' learning. The video cases produced for this online course feature two Year 8 students, Tim and Tom (pseudonyms), both age 12, attending two different secondary schools in a large city in the United Kingdom. Since what the students did with the digital environment provided was of importance and relevance, a screencast video-recording software was used to enable video recording of students' on-screen work as well as an audio recording of any student-student interactions while working through the mathematics activity.

10.3.3 Description of Our Video Cases

An overview of the four short videos are shown in Table 10.2. Figure 10.2 is an example of what the video cases look like, together with some explanations of the areas of the video screen that participants should pay particular attention to. Figure 10.2 shows the boys (video capture of their faces) talk through the activities (audio recorded) as they use a digital environment to do some mathematics

Table 10.2 Description of videos

Video Case 1 (3 min)	Video Case 2 (8 min)	Video Case 3 Part A (1 min)	Video Case 3 Part B (6 min)
Straight line graphs	*More straight line graphs*	*Mid-points in a quadrilateral*	*Mid-points in a quadrilateral*
Plotting points *in a symbolic and graphical environment* that lie on straight lines of given equations	Finding the equations of straight line graphs already plotted *in a symbolic and graphical environment*	Recording of students' work while investigating the nature of the quadrilateral made by joining up the mid-points in a quadrilateral	Recording of students' work while investigating, *in a dynamic geometry environment, the* nature of the quadrilateral made by joining up the mid-points in a quadrilateral

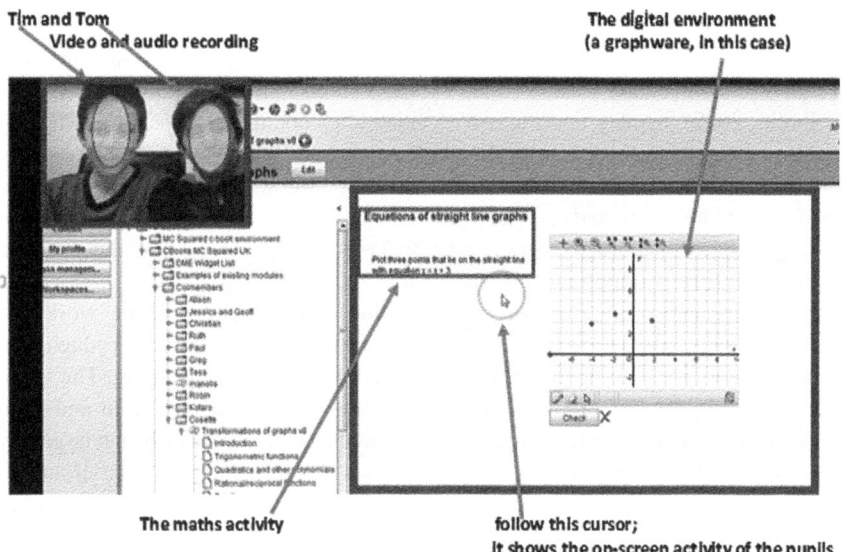

Fig. 10.2 A screen shot from Video Case 1

(their on-screen activity being captured, too). The boys were invited to work independently from a teacher. They were encouraged to talk through and to each other when working towards the solution to the mathematical activities they were presented with. Once the recordings were edited, the short video cases (not longer than 10 min each) were uploaded online.

The ethical dimension of creating and using these video cases was considered thoroughly. A review of some of the literature on the use of videos raised my awareness of the **ethical considerations** when images, video or audio recordings are taken, then posting them online (Flewitt, 2005). Permission to use the videos was sought through students and parents' consent, where my intention on how to use the video material and for what purposes was clearly explained.

10.3.4 Piloting the Use of Video Cases

During the second presentation of this online course in (Spring, 2016), the video cases were implemented for one of the three themes of the course, namely Theme A: Visualising. In this theme, which spread over weeks 1–4 of the course, the course participants explored the value of access to multiple representations enabled by the digital technology in terms of the potential to facilitate learners' understanding of various areas of mathematics. The participants were expected to familiarize themselves with multiple representation software (such as graphing packages and dynamic geometry software, thus developing their TK—technological knowledge and skills) and experience for themselves the potential and limitations of these applications (contributing to development of their TCK—knowledge and skills concerning the combination of mathematics and the use of technology) in facilitating learners' understanding of the concepts and properties associated with functions and their graphical and symbolic representations and in promoting spatial and geometrical reasoning in mathematics (hence informing their TPK—knowledge and skills concerning the combination of mathematics and didactics of mathematics) (Mishra & Koehler, 2006). Two weeks into studying for Theme A: Visualizing, for the end of theme task, the participants were asked to reflect critically on the implications of using such technology in the learning and teaching of mathematics. In designing the end of theme task, the participants were invited to strategically select particular sequences of the uploaded video cases that were significant to them and write their reflections on how the students' learning had been affected by doing mathematics in a digital environment. By choosing to focus on specific parts of the chosen video(s), the participants were invited to explain their new thinking and insights through engagement with research and ideas assimilated from the literature reviewed (the key course readings) in order to evaluate and justify the implications of using digital technology for the students' learning as portrayed by the video cases.

In my role as a tutor, I provided scaffolding for the participants by modelling engagement with research and theory when analyzing learning episodes of the video cases. For example, I exemplified how I selected episodes in one of the video cases, how I annotated the video to focus on specific aspects of students' interactions with the digital tools which were significant to conceptual understanding of the mathematics under scrutiny and how I then analyzed and interpreted students' learning, with annotations and explicit links to research and theory. A forum discussion of the tutor's writing followed, to further sensitivities and increase the participants' awareness of the analysis and interpretation of the learning and of the need to be explicit in making connections between research & theory reviewed and their observations of the students' learning/activities in these videos. They were made aware of the expectations of engaging with the literature reviewed to support and back up claims such as 'all students understood', 'ICT helped', 'all learned', etc. otherwise made by participants in our course in the previous year.

The participants were invited to watch the short video cases, then pause and reflect on how and in what ways the digital technology together with the mathematics tasks designed supported students' learning. Taking van den Berg (2001)'s suggestion into account, participants were guided to attend to the more sophisticated and less obvious aspects of doing mathematics through using the digital technology. In this respect, further scaffolding was provided to the participants through a number of guiding prompt questions to use while watching the videos: What would you consider as the benefits/limitations of using digital technology in this mathematics task, compared to a similar mathematics task but in a non-technology environment? What representations of this concept are facilitated by doing this activity in a digital environment? How did the students employed the digital environment to investigate the mathematics task and why? How did the design of the task support the students' consolidation of and extension of their knowledge about the mathematics concept/topic?

The participants' written accounts for the end of theme task were shared online, hence shifting the focus of their online communications about each other's accounts of the learning in the *shared* episodes from the 'what happened' to their analysis and interpretation of 'the how and the why' supported by their own engagement with the key readings of the course. In the online learning environment of this course, there were opportunities for asynchronous contributions from all the participants. They were encouraged to engage with and learn from each other's contributions by watching the significant episodes each of them selected and then read about each other's analysis and interpretation. They could then reflect at their own pace on how each of them used the ideas assimilated from the key readings together with their personal knowledge and experiences in order to evaluate and justify how the student's learning of mathematics benefitted from using digital technology. One participant's contribution to the forum discussion illustrates how she benefited from reading accounts of learning episodes that may be different to your own interpretation *"Thanks for your comments, Mark. I have also read your written work and I appreciated your suggestion that teachers might have made students explore different quadrilaterals and discuss about the new construct. That was a really good opportunity for us to watch the videos which simulate a real teaching situation and to identify key points about them. I too felt lucky being able to access my friends' opinions."* (Dina's contribution, week 4, forum discussion)

10.4 Participants and Data Sources

All the participants (16) on our second presentation of the online course have agreed for their written contributions to be used as data for this research study. They constituted a convenience sample for researching whether and how the pedagogical intervention supported and contributed to the development of the participants' RiTPACK, with a particular focus on how the digital environment supports students' mathematical work. Gray (2014) notes that research that "… tries to understand what is happening … explores the personal construction of the individual's world [and] studies individuals … using small samples researched in depth or over time" (p. 12). Miles et al. (2013) in Gray (2014, p. 174) advise the selection of information-rich cases which can be studied in depth. For this reason, in this chapter the qualitative data collected and analyzed for the purpose of the study reported in this chapter consisted of one participant' online contributions throughout Theme A of the course, his analysis and interpretation of the chosen episodes from the four video cases and his final assignment for this course; the assignment documented the participant's personal development of a mathematical idea or topic based on their exploration of digital technology and reflection on their experiences of designing and testing the use of the activity with learners.

10.5 Data Analysis

The data were analyzed using the RiTPACK lens. The conceptual framework structured what I noticed and paid attention to and took as important in the analysis of the data collected. My goal was to describe the development of the participants' TPACK components, with a particular focus on their learning about how digital environment supports students' mathematical work, and to find evidence of them engaging with the theoretical and research knowledge base of the course when analyzing and interpreting their accounts of students' learning.

Simon and Tzur (1999) talked extensively about *the generation of accounts of teachers' practice* as an attempt to understand teachers' current practices in a way that accounts for aspects of practice that are of theoretical importance, using conceptual frameworks developed in the research community. They characterized their methodology as "explaining the teacher's perspective from the researchers' perspectives" (ibid., p. 254) and it was developed as an alternative to both deficit studies where the principal focus is on what teachers' lack, do not know or are unable to do, and teachers' own accounts of their practice.

The RiTPACK lens also enabled me to identify in participants' written contributions explicit instances of where and how their analysis and interpretation of the mathematical learning was informed by theory and research.

Thus, the evaluation of the pedagogical intervention of this study consists in analyzing the developing 'quality' of a variety of the written contributions of the participants throughout the delivery of this course, where quality was evidenced in the levels of development of participants' engagement with the theoretical and research knowledge base of this course to analyze and interpret students' learning. In the following, I will be reporting on one participant (Mark—pseudonym)'s trajectory towards the development of his RiTPACK.

10.5.1   The Case of Mark

Prior to the start of the course, all the participants on this course were asked to submit a short piece of writing about the digital technology use in their own learning and teaching of mathematics. By sharing these writings online, the participants were thus encouraged to get to know each other's backgrounds and experiences with the new technology.

10.5.1.1   Week 1

Mark, an experienced mathematics teacher, expressed his own views about the potential of digital technology: *Much technology used inappropriately simply does the same thing as non-technology, but used well [it] has the ability to add significant value* (forum discussion, week 2), with no further exemplification of his claim. Prior to his enrolment on the course, Mark had invested into developing his TK (technology knowledge): *My own experiences with technology is that I have spent a considerable amount of time in developing my knowledge and getting to know systems, to the point that I would probably have got better student outcomes by doing something else* (forum discussion, week 1), and at the start of the course he expressed his hopes that *I am getting to the point of pay off.*

Mark's writing at this stage is descriptive, drawing from his own experience with digital technology prior to starting the course.

10.5.1.2   Week 2

In week 1 of the online course, the participants were introduced to some key readings aimed at raising their awareness of the TPACK literature. In a written task at the end of the week 2 of this course, the participants were asked to describe their own TPACK components, namely knowledge, skills, and experiences on using the digital technology in their own mathematics learning and in their teaching, by exemplifying them with specific instances from their practices. Like most of the other participants on this course, Mark did not illustrate any of the claims about the development of his TPACK components with specific examples from his own experience with digital technology or from his own classroom practice. Instead, his writing consisted of assertions about digital technology use in doing mathematics, without being clear if they were inferred from his practice or if they were just personal opinions, without empirical evidence. For example, Mark remarks that *Computer system is engaging. It allows participants to experience a variable by dynamically changing it and seeing the results "what is the same, what is different"* (forum contribution, MarkTPACKstory, week 2) which could otherwise be an indicator of his TCK. Referring specifically to his TPK, Mark envisaged his role in *show*[ing] *students what actually happens using dynamic functionality; instantaneous graphing and tabulating of results of expression allows for students to see the effect of a varying variable in these forms* (forum contribution, MarkTPACKstory, week 2). In his writing, there is evidence that his own awareness of how digital tools allow for the *interplay between representations dynamically* (an indicator of his TCK) influenced his view of how digital technologies could be used to support his students' learning *by seeing the same thing in different ways* and by *promoting thinking through questioning on predicting potential changes* (an indicator of his TPK) (forum contribution, MarkTPACKstory, week 2).

Mark's writing at this stage is a mix of descriptions and claims about some of the potential benefits and limitations of the digital technology for learning, but with no evidence that connects the claims to specific events from either his experience or his practice.

10.5.1.3   Week 3

For the following week of this course (week 3), the participants themselves explored the value of access to multiple representations in terms of the potential to facilitate students' understanding of various areas of mathematics. They were asked to use a piece of symbolic and graphical representation software to investigate how the parameters in the general form of a quadratic equation were related to the graphical representation of the equation and share reflections on their own learning experiences. In his online entry, Mark comments on the importance of and the need for creating *many images to construct relationships that will facilitate visualisation and reasoning. This is where the technology is powerful in facilitating the creation of many images rapidly in order to focus participants on the connections between them. Technology is also engaging and provides a change from the "normal"* (forum contribution, week 3). This is a big claim about the potential of digital technology, indicating his knowledge of TCK and TPK, but again, with no specific reference to the actual mathematics investigation he carried out, nor with an explicit insight into how it benefitted his own investigation of the task. Similarly, when asked to summarize his reflections on the learning opportunities facilitated by the use of a dynamic geometry software, Mark's writing provides evidence of his engagement with the key course readings (RiT): *The added value from the dynamic nature is how variance can be shown and more complex mental images can be created in participants' minds since they will see multiple images of the same problem. This can only enhance participants understanding and engagement* (*from* Laborde, 2005) (forum contribution, week 3), but he fails to link the research knowledge base of the course with his own experience when using the dynamic geometry software. At this stage, there is evidence that Mark's writing is descriptive, with some attempts to draw on the key readings, but this is not done explicitly.

While I wanted Mark and the other participants to continue to engage with research through using the ideas assimilated from the literature reviewed, I wanted to support them in noticing and interpreting students' learning when doing mathematics in a digital technology environment, by focusing on not only on 'what is actually happening' but also on 'how and why'. The video cases were introduced and tutor's modelling of analysis of an episode of students' learning was shared with the participants.

10.5.1.4   Week 4

For the end of theme task, Mark selected an episode from a video showing Tim and Tom working together to find the equation of two straight line graphs intersecting each other at a point. The significant episode he selected 'starts' at the point when the boys typed in a partially correct but incomplete equation of one of the two straight line graphs. Mark comments on how the feedback from the dynamic software *exposed* [the boys] *to a misconception when the technology shows them the graph of y = 4x* [which] *is different from the graph they are trying to define. Here they are able to quickly alter their incorrect conjecture as a result of timely response from the technology. Additionally, rather than just being told they are wrong and, as a result of the technology showing them the graph of their conjectured function* [the inputted equation] *beside the target function, they see that the coefficient of x is related to the steepness* [of the straight line graph]. *They both alter their conjecture fluidly and add clarity to their visualisation of the situation.* Mariottii and Pesci (1994) *cited in* Elliot (1998) *say that visualisation occurs when 'thinking is spontaneously accompanied and supported by images'* (End of Theme A task, week 4). I see here a detailed description of the learning episode selected. Mark explains what the boys are doing, at the same time connecting his interpretation of the boys' actions with research and literature in an attempt to justify his evaluation of how the boys' learning benefitted from using the digital environment (an indicator of his RiTPK). Mark goes on to notice that the boys *add another image to the "family" of images.* Through doing so, *this connection between the coefficient of x and the gradient is again confirmed when their next conjecture of y = 2x − 4 turns out to be too steep again, so they correctly reason that they need to reduce the coefficient of x again* (End of Theme A task, week 4). When analyzing this observation of pupils' actions, Mark draws on his PCK about pupils' learning of this topic, which he then links to the specificity of the digital technology environment by explicitly making connections to Solano and Presmeg (1995) *cited in* Elliot (1998) [who] *see visualisation as 'the relationship between images'* to explain the boys' actions of using the software to sketch straight line graphs of equations inputted by them and improve their equations based on the feedback from the software (an indicator of his RiTPACK). He then explains how each time the feedback scaffolds the boys' learning *in order to visualise there is a need to create many images to construct relationships that will facilitate visualisation and reasoning* and concludes that the boys did benefit from the digital environment as *in this thinking process another image is added to their visual understanding and they gain further clarity* (End of Theme A task, week 4).

10.5.1.5   End of Course Assignment

In his end of course assignment, Mark describes one of his students' work with a dynamic geometry software: *Student 2, at the end of Task 3* [which Mark designed for his final assignment], *when asked about his understanding of Thales theorem said "I can actually see it". This implies that during the tasks he gained a clear visual picture of the Theorem, which he did not have before. The student's reference to being able to "see" the Theorem seems to link closely with the research on visualisation for understanding* (End of course assignment). In his assignment, Mark's RiTPACK is made visible through his explanation of how his review of the literature on visualization influenced his design of the student Task 3: *In Geometric Visualisations, visualisation is when students can perceive a family of images with the same "geometric make-up"* (Healy & Hoyles, 2000). *The ability to make connections between images facilitates reasoning and is therefore critical in forming and proving new mathematical ideas that could later become theorems once proven. Visual methods of solution complement and provide an alternative to a traditional symbolic approach used in mathematics* (Zimmermann & Cunningham in 1991).

At this stage, Mark's writing is concerned with analysis and interpretation by drawing consistently on the literature and research and he even begins to offer pedagogical solutions based on his interpretations: *This suggests that students will benefit from approaching a problem in both a visual and traditional symbolic way and each will add something to the students' understanding* (End of course assignment) providing further evidence of the development of Mark's RiTPACK.

10.6   Implications and Conclusions

The primary focus in this research study was the development of one of the four aspects of the participants' RiTPACK, namely their knowledge of students' learning with technology, through a pedagogical intervention. The analysis of Mark's written contributions over the first four weeks of this course indicated that Mark's RiTPACK was developing. While there is evidence that Mark started developing his TPACK and engaged with research right from the start of the course, the connection between these two aspects of his learning on the course was not established until later in the course (weeks 3 and 4). There is evidence in his end of course assignment that the his prior engagement with the video cases (the peda-gogical intervention in week 4) supported Mark in writing about and reflecting on specific instances where the digital technology supported students' thinking about and learning of mathematics, which he analysed and interpreted through engaging with the key readings (Ri) and connecting it with his personal knowledge and experience (TPACK).

Through this pedagogical intervention, the intention was to support Mark (as well as all the other participants on the course) become more actively engaged with the research and knowledge base of this course rather than just ingurgitating messages that 'experts' put forward/proclaim about the potential of digital technology.

From the work presented here, I propose that through using video cases teacher educators could support participants in Masters level courses learn how to critically analyze practice. This is significant for several reasons. Firstly, the designed intervention was brief, consisting of an intervention early on in the delivery of the course (week 4), at a time when the participants have started developing their TPACK (specific to Theme A) and started engaging with the key readings of the course. Reflections on previous presentations of the course provided clear evidence that the participants found it challenging to apply the ideas encountered in the key readings when reflecting on students' learning with digital technology. Through tutor modelling on how to engage with research and theory when analyzing learning episodes of the video cases, participants' awareness of actively engaging with theory and research increased and supported them in how to make this explicit in their writing. This was an important aspect of the intervention, as in an online course writing is the only means of communication when teaching and in peer collaboration.

Secondly, the video cases provided the participants with shared learning epi-sodes to analyse which together with sharing their written accounts supported further the participants in critically engaging with (different interpretations) of 'the how and the why' and where each participant's analysis and interpretation was supported by the research and theory reviewed.

Another dimension of this research study was the advance of the RiTPACK theoretical framework. As exemplified through Mark's case study, RiTPACK framework can provide an analytical and yet pragmatic tool in supporting teacher educators raise the critical awareness needed for teachers to reflect on their practices.

References

Calandra, B., Brantley-Dias, L., & Lee, J. K. (2009). Using video editing to cultivate novice teachers' practice. *Journal of Research on Technology in Education, 42*(1), 73–94.

Elliott, S. (1998). Visualisation and using technology in A Level mathematics. In E. Bills (Ed.) Proceedings of the BSRLM Day Conference at the University of Birmingham (pp. 45–50). Coventry: University of Warwick.

Flewitt, R. (2005). Conducting research with young children: some ethical considerations. *Early Child Development and Care, 175*(6), 553–565.

Gray, D. E. (2014). *Doing Research in the Real World.* London: SAGE.

Healy, L., & Hoyles, C. (2000). A study of proof conceptions in Algebra. *Journal for Research in Mathematics Education, 31*(4), 396–428.

Koehler, M., & Mishra, P. (2009). What is technological pedagogical content knowledge? *Contemporary Issues in Technology and Teacher Education, 9*(1), 60–70.

Laborde, C. (2005). Robust and soft constructions: Two sides of the use of dynamic geometry environments. In H.-C. L.-C. Sung-Chi Chu (Eds.) *Proceedings of the Tenth Asian Technology Conference in Mathematics,* (pp. 22–35). South Korea: Advanced Technology Council in Mathematics, Incorporated.

Leat, D., Lofthouse, R., & Reid, A. (2014). *Chapter 7: Teachers' views: Perspectives on research engagement.* Research and Teacher Education, The BERA-RSA Inquiry.

Liebenberg, L. (2009). The visual image as discussion point: Increasing validity in boundary crossing research. *Qualitative Research, 9*(4), 441–467.

Mishra, P., & Koehler, M. J. (2006). Technological pedagogical content knowledge: A new framework for teacher knowledge. *Teachers College Record, 108*(6), 1017–1054.

Otrel-Cass, K., Khoo, E., & Cowie, B. (2012). Scaffolding with and through videos: An example of ICT-TPACK. *Contemporary Issues in Technology and Teacher Education, 12*(4), 369–390.

Pepin, B., Gueudet, G., & Trouche, L. (2013). Re-sourcing teacher work and interaction: New perspectives on resource design, use and teacher collaboration. *ZDM: The International Journal of Mathematics Education, 45*(7), 929–943.

Simon, M. A., & Tsur, R. (1999). Explicating the teachers' perspective from the researchers' perspective: Generating accounts of mathematics teachers' practice. *Journal for Research in Mathematics Education, 30*(3), 252–264.

Solano, A., & Presmeg, N. C. (1995). Visualization as a relation of images. In L. Meira, & D. Carraher, (Eds.), In *Proceedings of the 19th PME International Conference, 3,* 66–73.

Van den Berg, E. (2001). An exploration of the use of multimedia cases as a reflective tool in teacher education. *Research in Science Education, 31.* Online: http://dx.doi.org/10.1023/A: 1013193111324.

Van Es, E., & Sherin, M. (2002). Learning to notice: Scaffolding new teachers' interpretations of classroom interactions. *Journal of Technology and Teacher Education, 10*(4), 571–596.

Yadav, A., & Koehler, M. (2007). The role of epistemological beliefs in preservice teachers' interpretation of video cases of early-grade literacy instruction. *Journal of Technology and Teacher Education, 15*(3), 335–361.

Zimmermann, W., & Cunningham, S. (1991). Visualization in Teaching and Learning Mathematics. *Mathematical Association of America, Washington, DC,* 1–8.

Part IV
MOOC and Rich Media Platform for Mathematics Teacher Education

Chapter 11
Design and Impact of MOOCs for Mathematics Teachers

Tamar Avineri, Hollylynne S. Lee, Dung Tran, Jennifer N. Lovett and Theresa Gibson

Abstract With online learning becoming a more viable and attractive option for students and teachers around the world, we discuss how one effort in the U.S. is focused on designing, implementing, and evaluating MOOCs designed for professional development of mathematics teachers. We share design principles and learning opportunities, as well as discuss specific impacts participants report for changes to teaching practices and how these MOOCs have impacted engagement of educators.

Keywords Online professional development · MOOC · Massive open online course · Mathematics teachers · Professional development design
Impact on practice

T. Avineri (✉)
North Carolina School of Science and Mathematics, Durham, NC, USA
e-mail: avineri@ncssm.edu

H. S. Lee · T. Gibson
North Carolina State University, Raleigh, NC, USA
e-mail: hstohl@ncsu.edu

T. Gibson
e-mail: tgibson2@ncsu.edu

D. Tran
Australian Catholic University, Melbourne, Australia
e-mail: dung.tran@acu.edu

J. N. Lovett
Middle Tennessee State University, Murfreesboro, TN, USA
e-mail: jennifer.lovett@mtsu.edu

D. Tran
Hue University of Education, Hue, Vietnam

© Springer International Publishing AG, part of Springer Nature 2018
J. Silverman and V. Hoyos (eds.), *Distance Learning, E-Learning and Blended Learning in Mathematics Education*, ICME-13 Monographs,
https://doi.org/10.1007/978-3-319-90790-1_11

11.1 Introduction

Improving mathematics and statistics instruction continues to receive attention around the globe, and many efforts have been made to design professional development for teachers to develop their content and pedagogy, typically on a small, local scale (cf. Darling-Hammond, Wei, Andree, Richardson, & Orphanos, 2009). Online courses can expand the reach of professional development and the teachers involved, fostering communities beyond school or district lines. Indeed, with advances in technology and interest in offering alternatives to traditional professional development (e.g., in-person conferences, workshops), the number of online professional development opportunities has increased. The National Research Council (NRC, 2007) claimed that:

> The provision of professional development through online media has had a significant influence on the professional lives of a growing number of teachers. Growing numbers of educators contend that online teacher professional development (OTPD) has the potential to enhance and even transform teachers' effectiveness in their classrooms and over the course of their careers. (p. 2)

Most recently, with increased demand for open, accessible resources and advances in technological and analytic capabilities, Massive Open Online Courses (MOOCs) have become a significant option for online education internationally (Pappano, 2012). MOOCs are designed and delivered in a variety of ways, depending on the learning goals for participants, to serve different target populations and provide diverse experiences for learners (Clark, 2013). Most MOOC participants engage in isolation, reviewing material individually and *perhaps* engaging in discussion forums (Kim, 2015). In recognizing the potential for MOOCs to serve as large-scale professional development, we are part of teams that have created MOOCs for Educators (MOOC-Eds) to assist mathematics and statistics teachers in developing content understanding and pedagogical strategies for improving practice, and forming local and global communities of educators. While MOOCs designed for educators have not had the "massive", large-scale enrollment of other MOOCs, they do reach larger numbers of people than typical online courses. MOOC-Eds are intended to attract professional educators who are specifically looking to engage in a free, open online course that is marketed to educators beyond specific geographical boundaries. Thus, a MOOC-Ed is a particular type of online professional development course that takes advantage of online technologies for learning that can connect educators across the world. Our question guiding this design and research is *"To what extent does a MOOC-Ed offer opportunities for mathematics and statistics teachers to engage in online professional learning and impact their practices?"*

11.2 Research on Design of Professional Development

In response to a growing focus on professional development for teachers, the American Federation of Teachers (AFT) published their *Principles of Professional Development: AFT Guidelines for Creating Professional Development Programs That Make a Difference* in 2008. They argued that:

> The nation can adopt rigorous standards, set forth a visionary scenario, compile the best research about how students learn, change textbooks and assessment, promote teaching strategies that have been successful with a wide range of students, and change all the other elements involved in systemic reform–but without professional development, school reform and improved achievement for all students will not happen. (p. 1)

In designing our MOOC-Eds to provide quality professional development (PD) for mathematics teachers that is effective in impacting their practice, we based many of our decisions on research on such PD in both traditional and online forms.

11.2.1 Mathematics Professional Development

Members of both the education research and practitioner communities have advocated for teacher education programs and professional development focused on the specific types of mathematical knowledge for teaching (MKT) and argue that providing such opportunities in a mathematics context can improve teacher quality (e.g., Conference Board of the Mathematical Sciences [CBMS], 2012; Hill et al., 2008). Without proper and frequent opportunities to engage in mathematics or mathematics education-specific PD, however, deepening of both mathematical knowledge and that knowledge for teaching would prove challenging. In an effort to address current challenges in mathematics education, the National Center for Education Evaluation and Regional Assistance (NCEE) of the Institute of Education Sciences (IES) at the U.S. Department of Education (Siegler et al., 2010) used empirical evidence gathered from a comprehensive collection of studies to develop specific recommendations for providing effective mathematics PD. They recommend that PD should:

- include activities that require participants to solve problems, explore the meaning and justification for algorithms, and discuss challenges associated with solving those problems;
- prepare teachers to use varied pictorial and concrete representations by including activities in which they develop tasks for their students that integrate these representations;
- develop teachers' ability to assess students' understandings and misunderstandings by incorporating research on teaching and learning as well as activities designed around critical analysis of student thinking (e.g., discussion of students' written work and/or video segments) (Siegler et al., 2010, pp. 43–44).

The CBMS (2012) also recommended that PD for classroom teachers be both content-focused and "directly relevant to the work of teaching mathematics" (p. 32). They provided ideas for social PD activities that bring mathematics specialists and other content experts together with practitioners to help strengthen teachers' MKT, such as solving problems and deeply exploring the mathematics in a professional learning community (i.e. fellow practitioners), analyzing authentic student work (e.g., from participating teachers' classrooms), and participating in collaborative task design with colleagues (CBMS, 2012). It was important to us as MOOC-Ed designers to incorporate elements that addressed each of these recommendations.

11.2.2 Online Professional Development

Recent research on online professional development (OPD) has identified design principles that help inform the development and implementation of future OPD opportunities that can have an impact on participants.

11.2.2.1 Personalization and Choice

Researchers have found that OPD that addresses the varied needs and abilities of its participants can be effective in changing teachers' instructional practice (e.g., Renninger, Cai, Lewis, Adams, & Ernst, 2011; Yang & Liu, 2004). One of the objectives of OPD programs should be to provide enough choice, or personalization, among the included activities (e.g., varied tasks, opportunities for reflection on different practices) to give participants options to engage with components that are most useful to them and their practice (Ginsburg, Gray, & Levin, 2004). Indeed, Renninger et al.'s (2011) findings suggested that "the potential of [OPD] lies in its designers' abilities to support participant stake by providing for multiple ways into thinking and working with disciplinary content—design that both accommodates and supports those with differing strengths and needs" (p. 229). Designers of OPD should be especially mindful of the activities they include in their program, as meaningful, accessible and relevant tasks encourage participants to then apply their knowledge to the classroom (Vrasidas & Zembylas, 2004). In providing choice and personalization, designers should "involve [participants] in the development of materials, so that online tools reflect what [they] want and need" (NRC, 2007, p. 20).

11.2.2.2 Online Communities of Practice (CoP)

Another principle that parallels one described in the context of face-to-face PD is the development and facilitation of an online community of practice (CoP). Researchers have highlighted benefits of such communities that are not always afforded in traditional face-to-face PD. For example, Mackey and Evans (2011) argued that online CoPs provide members with "extended access to resources and expertise beyond the immediate school environment" (p. 11), thereby offering ongoing PD and the potential for increased application of learning into the classroom. In their study on teacher OPD exploring the "impact and transfer of knowledge" (p. 191), Herrington, Herrington, Hoban, and Reid (2009) found that teachers succeeded in implementing new pedagogical strategies in their classrooms when they felt supported by their online CoP.

In order to maximize the benefits that CoPs provide, designers of OPD programs must be creative in building the infrastructure necessary to support such communities, as participants have the challenge of not being physically in the same place when engaging in online activities. Programs that include asynchronous discussion forums where participants respond to carefully-designed, relevant prompts provide opportunities for participants to reflect on their practice, exchange ideas, and discuss strategies for improvement on their own schedules and with colleagues they may not interact with otherwise (Treacy, Kleiman, & Peterson, 2002).

All of this suggests that opportunities for interaction and discussion offer participants the opportunity to engage in learning that will be sustained and relevant, as knowledge is enhanced through the exchange of thoughts and insights among a CoP, and skills are developed with a focus on specific needs (Simoneau, 2007). In the next section, we describe how we designed our courses to align with the literature on PD for mathematics teachers.

11.3 Design of Courses

The MOOC-Ed effort at the Friday Institute for Educational Innovation (http://www.mooc-ed.org) includes a collection of courses built using research-based design principles of effective professional development and online learning (Kleiman, Wolf, & Frye, 2014) that emphasize: (a) self-directed learning, (b) peer-supported learning, (c) job-connected learning, and (d) learning from multiple voices. The focus of this research was on two of the MOOC-Eds on mathematics and statistics education, namely *Fraction Foundations* (*FF*) and *Teaching Statistics through Data Investigations* (*TSDI*). The *FF* MOOC-Ed was designed to help K–5 teachers teach fraction concepts (selected from K–5 USA curricula) and skills through understanding students' thinking and implementing research-based approaches in classrooms aligned with recommendations of a practice guide for developing effective fraction instruction (Siegler et al., 2010). The purpose of the *TSDI* course was for participants to think about statistics teaching in ways likely different from current practices in middle school through introductory statistics. A major goal was for teachers to view statistics as an investigative process (pose, collect, analyze, interpret) that incorporates statistical habits of mind and view learning statistics from a developmental perspective, aligned with guidelines from Franklin et al. (2007). In the following sections, we provide examples to illustrate how we enacted the four design principles in these two courses.

11.3.1   Self-directed Learning

We promoted *self-directed learning* by encouraging participants to set their personal learning goals at the beginning to help guide their experience. For example, in the *FF* course, participants engaged with approximately 30 items to consider their awareness of various mathematical and pedagogical issues in teaching fractions; in the *TSDI* course, participants engaged with a confidence rating survey for their statistics teaching efficacy. In each course, multiple types of resources were incorporated, such as classroom-ready materials (e.g., lesson plans, tasks, instructional videos) and thought-provoking materials for educators to reflect on their practice and deepen their content and pedagogy knowledge for teaching. These resources were often provided with multiple media, such as readings/transcripts, classroom videos, animated videos, and podcasts to support different paths of learning activities. Participants were also often given choices to explore materials designed for different levels of understandings or grade levels. In addition, the project component of each course was designed to suit a teacher's practice or to assist other educators, such as mathematics coaches, professional development providers, and mathematics teacher educators, in developing their own materials for use in professional development settings.

11.3.2   Peer-Supported Learning

One way that MOOC-Eds differ from other MOOCs is the intentional design to create and support a *network of professional learners*. This is accomplished through opportunities to interact with one another in online discussion forums, review each other's projects, rate posted ideas, recommend resources, crowdsource lessons learned, and participate in twitter chats (Kleiman & Wolf, 2016). The discussion forums were designed for participants to post their thoughts about videos and discussion prompts, as well as interact with others, including facilitators of the courses. It was important for us as OPD designers to construct thoughtful prompts and questions that were open-ended to a degree that motivated participants to engage in discussions that were relevant to them. The design teams functioned as facilitators in forums; we encouraged participants to share experiences, asking probing questions to elicit participant thinking, challenged current understandings, and connected related discussion threads from different groups to offer multiple perspectives to support richer discussions. Our role as facilitators included sending weekly announcements to all enrolled and engaging in discussions along with participants in the discussion forums. This again makes a MOOC-Ed different from other types of MOOCs; in a MOOC-Ed, the instructors/facilitators have a presence in communications beyond simply answering support questions.

11.3.3 Job-Connected Learning

Through the use of video case studies, activities and projects assigned to develop classroom materials, both courses provided opportunities for participants to learn lessons that could enhance their current practice and was *directly related to their work*. Video lessons, student interviews, small group teacher videos, and resource videos provided authentic images of students working through problems, revealing their thinking, and models of questioning strategies. Such resources provided opportunities for participants to think deeply about their approach and consider alternative strategies. Expert panel videos focused discussions from experienced educators around anticipated student difficulties and effective teaching approaches.

11.3.4 Learning from Multiple Voices

Both courses incorporated a number of opportunities to *learn from multiple voices*. As members of the design teams, we created many of our own resources, but also used existing open access resources written by other educators in the respective disciplines. Discussions that included well-known experts in each discipline were recorded and used throughout the courses. In these videos, the experts discuss relevant topics, share personal experiences and valued resources, and suggest strategies for implementing knowledge gained from research in everyday classrooms. Teacher and student voices were brought into the courses through videos of teachers implementing investigations in real classrooms, including teacher commentary. These open-access videos, all available online, gave participants a look inside real classrooms to imagine possibilities for their own contexts.

Student voices were brought into the courses through student interviews edited to highlight specific student thinking (see Fig. 11.1a) and through brief animated videos constructed based on actual student responses from research (Fig. 11.1b). The student videos were used to: (1) present purposeful aspects of students' work while they solved tasks; and (2) ask participants to specifically consider students' work using frameworks and constructs introduced in the course and how they might engage their own students in such tasks.

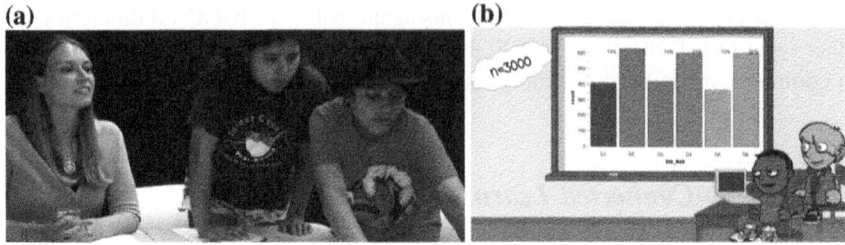

Fig. 11.1 **a** A group interview of students working on a task, and **b** animated video of students' work using technology to investigate a task

11.4 Framing Learning Opportunities in the MOOC-Eds

Both courses were built upon research-based recommendations for effective teaching in the specific content areas. All resources included in the courses centered around and explicated the recommendations through short videos, documents and experiences that are applicable for educators. In the *FF* course, many activities were built using the recommendations in the Siegler et al. (2010) report on effective instruction on fractions. For example, in a unit about using fair sharing activities to build on students' intuitive understandings and support understanding about fractions, the *What Would You Do Next?* videos of students working on fair sharing tasks with an interviewer prompt for participants to examine the students' understanding and how they would help move students forward. Core resources included research-based recommendations for using fair sharing activities in teaching fractions and other resources to support the teaching of concepts focusing on student thinking. Discussion forums enabled participants to focus on personal reflection on their successes as well as challenges when teaching specific fraction constructs.

In the *TSDI* course, research on students and teachers' learning of statistics and teaching practices was used to build opportunities for engagement. For example, we built upon an existing framework (GAISE, Franklin et al., 2007) by incorporating recent research on students' statistical thinking and highlighting productive statistical habits of mind. The new framework, Students' Approaches to Statistical Investigations [SASI], needed a variety of learning materials and opportunities for participants to develop an understanding of its importance and potential ways it can influence their classroom practices. Both a static and interactive version of a diagram was created to communicate the investigative cycle, reasoning in each phase at three levels, and an indication of productive habits of mind for each phase (Fig. 11.2).

Two brief PDF documents described statistical habits of mind and the framework. In a video, the instructor illustrated the framework using student work from research, and another video featured one of the experts illustrating the development of the concept of *mean* across levels of sophistication. The participants could also watch two animated illustrations of students' work on a task that highlighted how

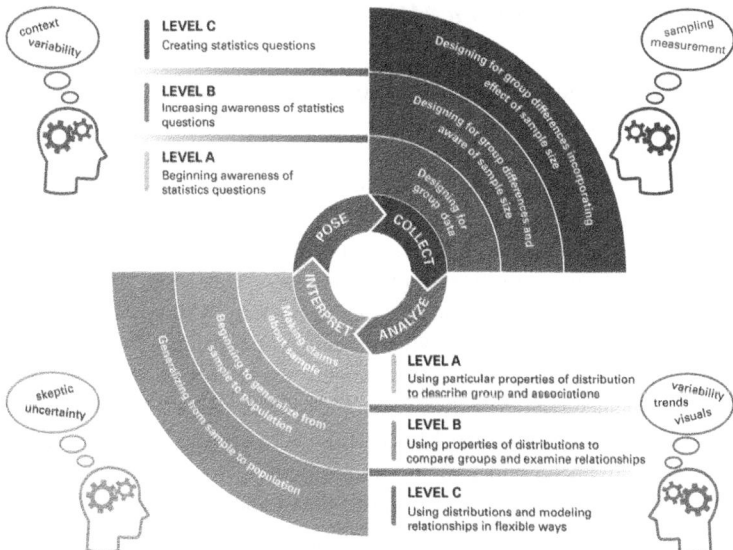

Fig. 11.2 Diagram of framework used in the *TSDI* course. Interactive version available at https://s3.amazonaws.com/fi-courses/tsdi/sasi_framework/index.html

students could approach an investigative task using different levels of statistical sophistication and then discuss, in the forums, students' work and how they could use such a task in their own practice.

11.5 Data Collection and Analysis

Although both courses have been offered more than once, we focus here on the course offerings of *FF* and *TSDI* that occurred in Spring 2015. Multiple sources of data were collected in quantitative and qualitative forms. This included course registration, pre-survey responses, and click logs of every action taken by participants (e.g., resources viewed, videos watched). All dialogs generated in discussion forums, 5-star ratings of resources, and feedback surveys for each unit and at the end of each course were collected. Descriptive statistics were generated based on demographic information, survey responses, and click logs. Open coding of forums and survey responses was used to develop themes related to participants' self-reported impacts on practice, as discussed below. In addition, follow-up interviews and classroom observations were used to ascertain impacts of the learning opportunities in the *FF* course on teachers' classroom practices.

11.5.1 Engagement in the MOOCs

There were 1712 participants registered for the *FF* course, with 34 countries represented (vast majority were USA based, 93%) while the *TSDI* course had 797 participants registered from 43 countries, with 597 (76%) registrants from the USA. Figure 11.3 illustrates our global reach for both courses in Spring 2015. In the figure, levels of participation are illustrated by the depth of color depicted on regions on the map. Darker regions represent higher levels of participation.

In both courses, the majority of participants were classroom teachers (58% *FF* and 64% *TSDI*). The courses also had approximately 10% of participants who worked in mathematics teacher education in university settings or other professional development roles. Interestingly, approximately two thirds of participants in both courses held advanced degrees (masters or doctoral). This is one indicator that many participants attracted to the MOOC-Eds were engaged learners in their discipline, valuing advanced educational opportunities.

The graph in Fig. 11.4 illustrates the engagement of those who enrolled. Participants were considered "no shows" if they never entered the course after registration, and were tagged as a "visitor" if they logged into the course and engaged with some aspect of it four or fewer times (Kleiman, Kellogg, & Booth, 2015). The remaining participants who engaged were considered active participants. As the graph indicates, there was a large proportion of registrants (1/4–1/3) that did not engage at any time. However, in both courses, there was a large number that either engaged somewhat or more fully.

While these numbers may not look impressive for a massively scaled course, the materials are potentially impacting a large number of participants in the context of professional development. Additional analysis was performed by Kleiman et al. (2015) that further characterized the active participants according to how engaged they were with videos, resources and tools, visiting discussion forums, posting in

Fig. 11.3 Global enrollment in both MOOC-Ed courses Spring 2015

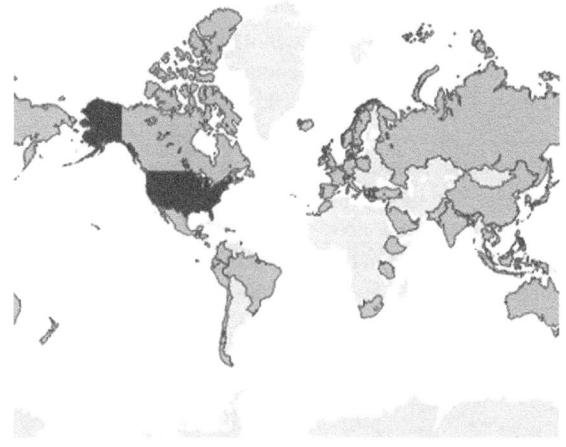

Fig. 11.4 Distribution of participation categories in both Spring 2015 courses

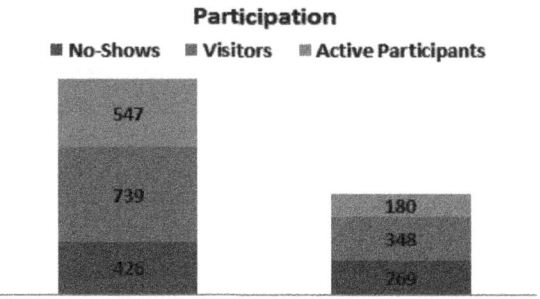

forums, and commenting on posts of others throughout the course. This analysis showed that the 547 active participants in the *FF* course were about equally categorized as declining activity (34%), moderate activity (38%), and sustained high activity (28%). However, the 180 active participants in the *TSDI* course had either declining activity (54%), or sustained high activity (46%). These high activity rates through the final units in the courses are much higher than typical MOOC completion rates (2–10%), but are aligned with completion rates when participants intend to complete a course (Reich, 2014).

We took a close look at click logs from the discussion forums to get a global sense of participants' engagement in the most publicly viewable form through forum posts and replies and comments to a peer or course facilitator. Across both courses, approximately 33% of participants (those classified as visitors or active participants) posted at least twice in the forums, with either a new post or comment on a post by someone else; however, Kleiman et al. (2015) identified many participants across both courses considered to be active contributors to the discussion forums. In *FF*, there were 770 participants (59.9% of 1286 visitors or active participants) who contributed in some way in the forums, with an average of 9.3 posts per participant. In *TSDI*, 308 (58.3% of the 528 visitors or active participants) participated in the forums with an average of 7.1 posts each. Examining click logs, we found there were also *many* more discussion views than postings. Some discussion views were done by participants who were active posters; however, other views were done by non-posters. Thus, many saw discussion forums as an opportunity for learning, even for only reading the posts of others. Participants who engage in such "lurking" are present, but not visible; thus, precisely why they read discussions and what they have learned from them is unknown.

11.5.2 Impact of the MOOC-Eds on Educators' Practices

On end-of-course surveys, participants were asked how effective they believed the MOOC-Ed was in preparing them to make positive changes in their practice. Across the two courses, 205 participants completed the optional end of course survey, with 96.5% of participants responding positively. Participants were then explicitly asked if they had made changes in their practice as a result of participation, to which 96.7% indicated "Yes." When asked to describe how they were applying what they learned to their professional practice, the following three themes emerged as the most common: (1) integrating new tools and strategies directly in their classroom; (2) implementing course projects or specific tasks/lessons; and (3) using course content for instructional coaching or professional development with colleagues.

For both courses, we examined discussion forum posts and open-ended responses on feedback surveys to look for explicit mention of changes to practice and tagged triggers for such changes. Preliminary analysis showed that there were several main themes and course triggers that participants indicated as leading to changes in practice.

In the *FF* course, the themes related to impact on practice were: (1) attending to student conception and misconception, (2) prompting students to elicit reasoning, (3) using multiple representations and models to help students understand fractions, (4) and designing/adapting rich tasks to assess students' understanding. These themes were associated with triggers from course experiences, such as the student interview videos, the tasks provided, and conducting a clinical interview as part of a project. Consider the following participant statements with course triggers bolded:

- I now really talk to the students and have **interviews** so I can assess better. I look at fractions in a whole different way. I also look closer at **student answers, I once would have considered incorrect**.
- The most valuable part of this MOOC-Ed was the **"What Would You Do Next?" video series**. As teachers, I think we need to see the "look fors" in students' misconceptions… For students to understand, teachers must become comfortable **seeing misconceptions** and addressing the understanding.
- I have begun to facilitate **learning of fractions versus teaching the students about fractions**. I am now having students take their time and **explore concepts in different ways** rather than rushing through and trying to teach an algorithm.

While some participants' comments addressed changes to their approach to teaching (e.g., increased focus on concepts as opposed to algorithms), others described how their participation supported their refined attention to and understanding of their students' thinking and their own personal improvement in knowledge of mathematics. This finding was supported by classroom observations and interviews with teachers who participated in the *FF* course (Avineri, 2016). Indeed, in describing her lesson planning after having participated in the MOOC-Ed, one teacher noted that the course encouraged her to:

> [try] to do something different on purpose. Trying not to drag out my old fraction folder and dig from it. Trying to do something other than I would normally have done…It has impacted me to just slow down. It's not a race. You have to strategically take your time and give them a chance to develop that knowledge, to start small and then let it blossom.

This was evident in observations of this teacher's classroom following the MOOC-Ed. Another teacher described how her enhanced mathematical understanding impacted her attention to students' thinking, which was also evident in classroom observations following the MOOC-Ed.

> I just wanted a better understanding of what a fraction was, what it meant to partition…and then to see [the students] today, [*speaking as a student*] 'Well you've got four more models, that's your four wholes that are represented, but you've got a fourth of each one or a third of each one'…I would never have looked for that conversation from them before.

In the *TSDI* course, four elements emerged as often cited for triggering impacts on practice. The most dominant trigger for change was the SASI framework. Not only did participants discuss how they needed to design tasks that met their students where they were, but also to further develop their levels of sophistication. Educators also reported wanting to use all four phases of a statistical investigation, rather than their past heavy emphasis on the analysis phase. Two additional common triggers were the use of technology for visualizing data, and use of real data sources that are multivariable and sometimes "messy". These triggers came from learning opportunities in the course that included videos of students and teachers engaged with messy data using technology, discussions in expert panel videos, and opportunities to use a large internet resource for gathering and sampling data from students across the world (*Census at School*) and a technology tool of their choice to engage with the data. Consider the following statements from participants with course triggers bolded:

- I have changed my planning process for statistics. I will use more **technology** in my teaching and spend more time on the **first 2 phases of the investigative cycle**. I will encourage **statistical habits of mind** and movement through the **levels of the SASI framework**.
- The **SASI framework** was the most useful part of the course. It is incredible. I've been telling the teachers here about it because normally we teach the Intro to Stats class only procedurally, **just calculations**, **with no sense of equations or interpreting**. But that has changed now because of using the framework.
- Since starting the class, I have had my students use **richer and messier data** in their investigations and I have also put **more of an emphasis on understanding the results** and being able to analyze findings.

While some of the comments indicate how teachers had already changed, or will change, their practices with their own students, other comments showed how elements of the *TSDI* course have impacted how participants encourage their colleagues to change their practices.

11.6   Discussion/Conclusion

The results are encouraging for how participants took advantage of our purposeful designs. There were many elements of these designs that acted as catalysts for self-reflection and change in practice. The use of frameworks and research-informed practices in teaching both fractions and statistics were highly valued and appear to assist participants in viewing the learning and teaching of these ideas more conceptually and comprehensively. Activities that were designed to capture the perspective of students, such as *What Would You Do Next?* videos, videos of students' work in the classroom, and edited and produced videos of students' work from research, all fostered rich discussions in the forums about students' understandings and how participants can use their new understanding of students' reasoning to inform instructional planning. Participants also seemed to be able to shift their perspectives from viewing the importance of teaching and learning these content ideas as reliant on algorithms or procedures, to a view of mathematics and statistics as more of a process that has nuanced conceptions that must be developed with extended experiences.

Too often, professional development is provided by local school districts and does not meet individual teacher's need (Sowder, 2007). However, our MOOC-Eds provided participants with the opportunity to engage in professional development to strengthen their content and pedagogy in areas of mathematics *they personally were interested in improving*. In the forums, one *TSDI* participant discussed the opportunities she felt the MOOC-Ed provided:

> Some all-day workshops can be painful and provide little benefit. I think teachers who have given up instructional time and been burned on a poorly designed workshop become increasingly resistant to later PD opportunities. This course has been just the opposite. I can engage with it on my own schedule, rather than losing class time, and I'm coming away with lots of new ideas, resources, and activities. I feel grateful for this opportunity and look forward to finding more like it.

The research-based design principles that guide the creation of these courses have afforded educators choice in professional learning, complemented with relevant, job-embedded activities, access to the perspectives of experts, teachers, and students, and a network of educators learning together around a common content area. For PD developers, our results can be used to help develop and refine future online PD opportunities for educators that are effective in having a meaningful impact on their practice. For school administrators, teachers, and PD providers, our results provide strong evidence that MOOC-Eds can serve as viable, cost-effective and quality online solutions to satisfy the urgent need for professional development for mathematics and statistics teachers (Avineri, 2016).

We continue to learn about the affordances and constraints of this model of professional learning for mathematics teachers and are interested in expanding our research. One expansion needed is to explore long term impacts on practice. There is also a need to explore the potential of MOOC-Eds beyond our current implementation model as time-bound courses managed primarily by our team. We

wonder, if the MOOC-Ed courses were available on-demand, how would participants engage, and would professional learning networks emerge without active
participation by facilitators? We would also like to explore the possibility of
international collaboration in the design of future courses, the impact of facilitators
in a course of this scale, and the possibility of offering smaller scale modules that
are continuously available.

There is much to be studied and understood about the effectiveness and impact
of MOOC-Eds as a medium for offering online professional development for
educators. Rich data emerged from our studies that can be further analyzed and
used to refine future MOOC-Eds, and more globally, online PD programs for
educators. This inspires motivation to continue the work toward making
MOOC-Eds and other OPD for educators effective in ways that serve educators
well in promoting changes in their knowledge and to have an actual, true impact on
their practice.

Acknowledgements The design, implementation, evaluation, and research of MOOC-Ed courses
at NC State University's Friday Institute for Educational Innovation is partially funded by the
William and Flora Hewlett Foundation. Any opinions, findings, and recommendations expressed
are those of the authors, and do not necessarily reflect the views of the Hewlett Foundation.

References

American Federation of Teachers (AFT). (2008). *Principles for professional development: AFT
guidelines for creating professional development programs that make a difference.*
Washington, DC: American Federation of Teachers, AFL-CIO.

Avineri, T. (2016). *Effectiveness of a mathematics education massive open online course as a
professional development opportunity for educators* (Unpublished doctoral dissertation). North
Carolina State University, Raleigh, NC.

Clark, D. (2013, April 16). *MOOCs: Taxonomy of 8 types of MOOC* [Web blog post]. Retrieved
from http://donaldclarkplanb.blogspot.co.uk/2013/04/moocs-taxonomy-of-8-types-of-mooc.
html.

Conference Board of the Mathematical Sciences. (2012). *The mathematical education of teachers
II.* Providence, RI, Washington, DC: American Mathematical Society and Mathematical
Association of America.

Darling-Hammond, L., Wei, R., Andree, A., Richardson, N., & Orphanos, S. (2009). *Professional
learning in the learning profession: A status report on teacher development in the United
States and abroad.* Dallas, TX: National Staff Development Council.

Franklin, C., et al. (2007). *Guidelines for assessment and instruction in statistics education
(GAISE) report: A pre-K–12 curriculum framework.* Alexandria, VA: American Statistical
Association. http://www.amstat.org/education/gaise.

Ginsburg, A., Gray, T., & Levin, D. (2004). *Online professional development for mathematics
teachers: A strategic analysis.* Washington, DC. Retrieved from http://www.eric.ed.gov.prox.
lib.ncsu.edu/contentdelivery/servlet/ERICServlet?accno=ED492927.

Herrington, A., Herrington, J., Hoban, G., & Reid, D. (2009). Transfer of online professional learning
to teachers' classroom practice. *Journal of Interactive Learning Research, 20*(2), 189–213.

Hill, H., Blunk, M., Charalambous, C., Lewis, J., Phelps, G., Sleep, L., et al. (2008). Mathematical
knowledge for teaching and the mathematical quality of instruction: An exploratory study.
Cognition and Instruction, 26(4), 430–511.

Kim, P. (Ed.). (2015). *Massive open online courses: The MOOC revolution*. New York, NY: Routledge.

Kleiman, G., Kellogg, S., & Booth, S. (2015). *MOOC-Ed evaluation final report*. Prepared for the William and Flora Hewlett Foundation. Raleigh, NC: Friday Institute of Educational Innovation. https://fi-courses.s3.amazonaws.com/place/research-reports/hewlett-evaluation-final.pdf.

Kleiman, G., & Wolf, M. A. (2016). Going to scale with online professional development: The Friday Institute MOOCs for Educators (MOOC-Ed) initiative. In C. Dede, A. Eisenkraft, K. Frumin, & A. Hartley (Eds.), *Teacher learning in the digital age: Online professional development in STEM education* (pp. 49–68). Cambridge, MA: Harvard Education Press.

Kleiman, G., Wolf, M. A., & Frye, D. (2014). Educating educators: Designing MOOCs for professional learning. In P. Kim (Ed.), *Massive open online courses: The MOOC revolution* (pp. 117–144). New York: Routledge.

Mackey, J., & Evans, T. (2011). Interconnecting networks of practice for professional learning. *International Review of Research in Open and Distance Learning, 12*(3), 1–18.

National Research Council. (2007). *Enhancing professional development for teachers: Potential uses of information technology, report of a workshop*. Washington, DC: The National Academies Press. Retrieved from http://www.nap.edu/catalog/11995/enhancing-professional-development-for-teachers-potential-uses-of-information-technology.

Pappano, L. (2012). The year of the MOOC. *The New York Times, 2*(12), 26–32.

Reich, J. (2014). MOOC completion and retention in the context of student intent. *Educause Review*. http://er.educause.edu/articles/2014/12/mooc-completion-and-retention-in-the-context-of-student-intent.

Renninger, K. A., Cai, M., Lewis, M. C., Adams, M. M., & Ernst, K. L. (2011). Motivation and learning in an online, unmoderated, mathematics workshop for teachers. *Educational Technology Research and Development, 59*(2), 229–247.

Siegler, R., Carpenter, T., Fennell, F., Geary, D., Lewis, J., Okamoto, Y., et al. (2010). *Developing effective fractions instruction for kindergarten through 8th grade: A practice guide* (NCEE #2010-4039). Washington, DC: National Center for Education Evaluation and Regional Assistance, Institute of Education Sciences, U.S. Department of Education. Retrieved from http://ies.ed.gov/ncee/wwc/PracticeGuide/15.

Simoneau, C. L. (2007). Communities of learning and cultures of thinking: The facilitator's role in the online professional development environment. In *Dissertation Abstracts International. Section A. The Humanities and Social Sciences*. Available from ProQuest Information & Learning database. Retrieved from http://ovidsp.ovid.com/ovidweb.cgi?T=JS&PAGE=reference&D=psyc6&NEWS=N&AN=2008-99110-216.

Sowder, J. (2007). The mathematical education and development of teachers. In Lester, F. K. (Ed.), *Second handbook of research on mathematics teaching and learning: A project of the National Council of Teacher of Mathematics* (Vol. 1, pp. 157–223). Charlotte, NC: Information Age Publishing.

Treacy, B., Kleiman, G., & Peterson, K. (2002). Successful online professional development. *Leading & Learning with Technology, 30*(1), 42–47.

Vrasidas, C., & Zembylas, M. (2004). Online professional development: Lessons from the field. *Education Training, 46*(6/7), 326–334.

Yang, S. C., & Liu, S. F. (2004). Case study of online workshop for the professional development of teachers. *Computers in Human Behavior, 20*(6), 733–761.

Chapter 12
Describing Curricular Materials for Mathematics Teacher Education in an Online, Rich Media Platform

Daniel Chazan, Patricio Herbst, Dana Grosser-Clarkson, Elizabeth Fleming, Janet Walkoe and Emina Alibegović

Abstract This chapter explores a way of describing the teacher education curricular materials being developed by mathematics teacher educators through their interaction with the Lesson*Sketch* online platform (www.lessonsketch.org). We briefly describe the platform and the larger project and then using the experiences created by two fellows illustrate the kinds of materials being created by the teacher educators. We then use Grossman's pedagogies of practice to explore how with the materials they are creating teacher educators are representing practice, decomposing it, and providing opportunities for their students to approximate practice through the curricular artifacts that they are creating. We note that the use of these pedagogies of practice to describe curricular artifacts suggests three different kinds of representation of practice in the artifacts being created.

Keywords Teacher education · Curricular materials · Practice-based teacher education · Online learning

D. Chazan (✉) · D. Grosser-Clarkson · E. Fleming · J. Walkoe
University of Maryland, 3942 Campus Drive, TLPL 2226C Benjamin Building, College Park, MD 20742, USA
e-mail: dchazan@umd.edu

D. Grosser-Clarkson
e-mail: dgrosser@umd.edu

E. Fleming
e-mail: fleming1@umd.edu

J. Walkoe
e-mail: Jwalkoe@umd.edu

P. Herbst
University of Michigan, Ann Arbor, MI, USA
e-mail: pgherbst@umich.edu

E. Alibegović
Rowland Hall School, Salt Lake City, UT, USA
e-mail: eminaalibegovic@rowlandhall.org

© Springer International Publishing AG, part of Springer Nature 2018   201
J. Silverman and V. Hoyos (eds.), *Distance Learning, E-Learning and Blended Learning in Mathematics Education*, ICME-13 Monographs,
https://doi.org/10.1007/978-3-319-90790-1_12

12.1 Introduction

Historically, the goals and purposes of teacher education have varied; for example, teacher education in the U.S. has shifted from a focus on particular teacher characteristics and behaviors to developing particular beliefs and forms of knowledge for teaching (Grossman & McDonald, 2008). At this moment in time in the U.S., against the backdrop of expected greater curricular uniformity across states (CCSSO, 2010), there is a call in teacher preparation for coordinating the content of teacher preparation across the country, for example by developing a common language and engaging in cross-institutional collective activity (McDonald, Kazemi, & Kavanagh, 2013; see http://corepracticeconsortium.com/). The field is considering the ideas of practice-based teacher education (for different perspectives, see Ball & Forzani, 2009; Zeichner, 2012), though even for those considering practice-based approaches, ways of describing the content of mathematics teacher education are not settled (e.g., consider the challenge of identifying high leverage practices as described by Ball, Sleep, Boerst, & Bass, 2009).

In the context of these developments in the curriculum of teacher education, this chapter explores the pedagogies of practice described by Grossman, Compton, Igra, Ronfeldt, Shahan, and Williamson (2009) as one way to describe interactive digital materials created to support practice-based teacher education. Specifically, the chapter uses Grossman et al.'s categories to examine the materials created by 12 Lesson*Sketch* Research and Development fellows. This examination suggests ways in which representing practice is a complex pedagogy, present in at least three distinct ways in which the fellows have created representations of practice.[1] Similarly, we note novel ways in which the online environment supports new opportunities for teacher candidates to practice the work of teaching. Before introducing the project and describing the materials created by the fellows, we use the mathematics education literature on curriculum to suggest that the curriculum creation process that is underway in teacher education, when it happens online, is influenced by the digital nature of technological artifacts (see Herbst et al., 2016).

12.2 Considerations in Specifying the Content of Teacher Education

Some readers may have had the experience of teaching both mathematics content courses and teaching methods (or didactics) courses for teacher candidates. The teaching of methods courses can seem quite different than the teaching of content

---

[1]Following Herbst, Chazan, Chieu, Milewski, Kosko, and Aaron (2016) (see especially p. 81), in this chapter, we will distinguish between representations as artifacts and the activity of representing practice.

courses. While mathematics is the content of a content course, it does not play the same role in a methods course; in fact, in a methods course, the role of content seems less easily specified.

While the process of transforming disciplinary activity in mathematics into a school subject matter has a long history (Popkewitz, 1987) and we are familiar with the results of this process, a similar process of transforming the practice of mathematics teaching into a curriculum for teacher education is also underway. In particular, the decomposition of the practice of teaching into component practices (as part of the move to practice-based teacher education) has the potential to privilege particular aspects of teaching. This process feels similar to the way that mathematics as a school subject privileges particular aspects of the discipline of mathematics.

New technologies are changing the nature of texts used in education settings. In particular, the move to digital platforms influences the curricular artifacts that hold the content of mathematics teacher education. For example, in digital environments, texts are more malleable and editable than they were in the past (think, e.g., of the difference between Wikipedia and the print encyclopedias that came before it) and this leads to questions about the boundaries of a textbook and who will author textbooks in the future (Chazan & Yerushalmy, 2014). New technologies are also creating opportunities for many new sorts of rich media educational texts (in the broader sense of text broached in Love & Pimm, 1996) that incorporate interactive diagrams (Yerushalmy, 2005), as well as video (e.g., Vi Hart's videos; https://www. youtube.com/user/Vihart) and animations (Chazan & Herbst, 2012). And, with the development of software platforms that facilitate online instructional interactions (e.g., Kahoot, Lesson*Sketch*, TedEd), rich media are being incorporated into formal education and put in service of particular curricular goals. Experiences for teacher candidates created with these platforms not only have pedagogical characteristics (Love & Pimm, 1996) in the sense that they put forward a way of learning about teaching, but also have curricular characteristics in the sense that those experiences also shape what it is that students, in this case teacher candidates, should learn. As practice-based teacher education is done through technological mediation (see Herbst et al., 2016), a transposition of the practice of mathematics teaching is being produced.

12.3 The Project and Platform Context that Informs This Chapter

Currently, as part of a National Science Foundation-funded project that explores the use of new technologies for the purposes of supporting practice-based teacher education, 12 U.S.-based mathematics teacher educators are each developing materials using the Lesson*Sketch* platform, where users can create, share, and discuss scenarios that represent classroom interaction (Herbst & Chieu, 2011). Each of these Lesson*Sketch* Research and Development (LR+D) Fellows has recruited an inquiry group of three to four other teacher educators to pilot and critique the materials that they have developed. These materials are being used in a range of teacher education contexts: methods courses for elementary or secondary teachers, mathematics content courses, and courses for in-service teachers. These materials are shared as Lesson*Sketch* experiences that teacher candidates complete, with those experiences including representations of classroom practice, as well as ancillary materials that function as teacher educators' guides to these materials.

Lesson*Sketch* currently has two kinds of users, advanced and basic. Both types of users can create depictions, or storyboards that are used to represent classroom activity with a graphic language made of cartoon characters. These representations can include mathematical inscriptions and diagrams as well as pictures of student work that can be placed on student papers or on the board of the depicted classroom. Speech in the classroom can be represented in text or in audio files, embedded in the depictions.

Advanced users also have access to a tool that allows them to author the agendas or scripts that the platform uses to run the experiences that advanced users set up for basic users. Advanced users can share those experiences with other teacher educators who can then modify the underlying agendas to customize the experiences for their teacher candidates (Herbst, Aaron, & Chieu, 2013). The platform supports the assignment of these experiences to students and allows teacher educators to review and provide feedback on what their students have done. The experiences students complete may include their annotation of video clips, their own creation of depictions, or their responses to prompts, all of which can then be annotated by their instructors using Lesson*Sketch*'s Annotate tool. These experiences, the agendas that create them, and the depictions that are used in them are the teacher education materials that the fellows are creating.

To summarize, one might describe the Lesson*Sketch* platform as an environment that provides teacher educators with editable experiences for teacher candidates, editable depictions of classroom interaction, and structures that support shared analysis of teacher candidate data. The LR+D fellows have been exploring the affordances of these capabilities for practice-based teacher education.

12.4 Two Experiences that Illustrate What Teacher Educators Have Created and How Their Teacher Candidates Interact with Experiences

In this section, focusing on the work of two of the fellows, Janet Walkoe and Emina Alibegović, we illustrate the sorts of experiences teacher educators have been creating for their teacher candidates and what it is like when teacher candidates interact with these materials.

12.4.1 Representing Student Thinking in a Methods Course

Fellow Janet Walkoe's goal is to allow teacher candidates to see and think about student thinking using information gathered in the classroom context as well as "behind the scenes" deeper insight into students' ideas. Central to the experience she has created, and a key resource in the materials she is creating, is a representation of teaching that teacher candidates are asked to analyze as homework to be done outside of class. In creating this representation of teaching for her class, Walkoe uses a blend of two different kinds of media; she uses student interview videos alongside depictions of classroom interaction (see Walkoe & Levin, 2018). For example, a classroom depiction displays students beginning to work on a proportional reasoning task in small groups; the task describes a pink paint mixture that is created by mixing three liters of red paint with four liters of white paint. When asked how to scale down the paint mixture to get one liter of pink paint, one student suggests mixing 3/4 L of red paint with 1/4 L of white paint. In the associated video, a cognitive interviewer delves into that student's way of thinking about the task by asking him questions that elicit and clarify his thinking. This array of media allows Walkoe to foreground skills involved in noticing student thinking.

Before beginning their interaction with the experience Walkoe has created, teacher candidates work on the paint-mixing task themselves. At the start of the experience, Walkoe uses the Media Show capacity of Lesson*Sketch* to show teacher candidates a classroom depiction in which a student, Raj, discusses the problem with his peers. Then, using the Question capacity of Lesson*Sketch*, teacher candidates are asked a series of open-ended questions, including what questions they as teachers would ask to probe Raj's thinking. Walkoe follows this by using Media Show again to have teacher candidates see how Raj thought about the problem in more detail, by watching a set of clips from a cognitive interview with Raj. The experience then returns to similar questions about Raj's thinking and asks teacher candidates how they might now want to interact with Raj in the classroom. Sequencing the tasks in the experience in this way supports teacher candidates in delving deeper into student thinking and using that analysis to inform questions they would ask the student in the classroom.

To illustrate how Walkoe's experience may help support teacher candidates' practice, we present the work of one teacher candidate. Donna, a teacher candidate enrolled in a mathematics methods course, interacted with the Lesson*Sketch* experience as part of a homework assignment for the methods course. Below we quote some of her written responses to prompts provided in the experience.

After viewing the initial depicted classroom scene, Donna wrote that Raj was struggling with ratios and fractions. When asked what questions she would want to ask him to probe his thinking, the types of questions she suggested were general, focusing primarily on *what* he was thinking about or *how* he obtained his initial answer. For example, Donna said she would ask Raj: "At first you thought the answer was 1/4 white and 3/4 red. How did you get your answer? Why did you change your answer to 3/4 white and 1/4 red? What were you thinking about? Can you explain how you created these fractions?"

The interview clip allowed Donna to see more detail in Raj's thinking and to realize that the richness of his ideas goes beyond what was initially seen in the classroom depiction. As she watched the first interview clip, her attention was drawn more to Raj's thinking and the details of his ideas, as intended. She was able to focus and articulate specific details about his thinking:

> When Raj talks about the numbers being 'too far apart in value' I suppose he is talking about 1/4 and 3/4. He then points to a number and says it is greater than another number and another number is greater than that. I'm not quite sure what he means by this. Maybe he is thinking about these numbers on a number line and the numbers aren't as close together as he originally thought. Maybe he is confusing their value or doesn't have the necessary number sense with fractions.

Donna is beginning to analyze his thinking, which is arguably a deeper way of attending to student thinking (Sherin & van Es, 2009) than the general way she discussed his thinking prior to watching the video clip.

When asked what questions she would want to ask Raj after watching the video clips, Donna suggested more specific questions, related to details about Raj's thinking:

> How did you get your original answer? What made you think your answer was wrong? Explain? Can you explain what you mean by 3L of red paint and 1L of white paint don't represent 3/4 and 1/4? Where did you get 3/4 and 1/4 from? What do you mean by 3/4 and 1/4 are too far apart? How does this affect your answer and reasoning? How do you know this value is greater than this value?

Seeing more detail in Raj's thinking supported Donna in crafting more specific questions to elicit Raj's thinking. This is the kind of work teachers do in classroom practice but have little chance to practice outside the classroom. Allowing details about student thinking to guide questioning pushes questions to have a different focus. For example, Donna's earlier questions focused on "why" or "how," while her later questions included "where did you get?" and "what do you mean by?" which have a different character.

Watching the interview also allowed Donna to see some of the conditions that helped elicit Raj's thinking. She commented, "I think once he was using a pencil and paper and making equations and not just thinking about the problem in his head, he remembered or realized the process to make the ratios proportionate." Donna also attended to the features of the interviewer's questions and prompts. "To elicit Raj's response the interviewer purposefully uses questions to elicit, probe, and connect Raj's mathematical ideas to his answer. This questioning supports his reasoning about this concept." Here Donna sees the direct relationship between the types of questions the interviewer asked and the detail in Raj's thinking the questioning elicited.

This LessonSketch experience allowed Donna to connect individual student's thinking to the larger classroom situation. She was able to see the richness of student thinking that lay behind the small piece visible in the classroom. When asked whether Raj's thinking in the interview surprised her, she commented, "This did surprise me. This was inconsistent with my ideas about what he would do [from the depiction]. I thought he would struggle with [the problem] more and need more prompting." The interview clips allowed Donna to see that the depth of student thinking went beyond the surface of what was visible in the classroom. These clips raise important questions for teacher candidates, like: How can I bring greater depths of student thinking into my classroom on a regular basis? and What do I do as a teacher if I cannot do cognitive interviews with every student in my class?

12.4.2 Representing Classroom Teaching in a Mathematics Course

Fellow Emina Alibegović also uses representations of teaching in her Lesson*Sketch* experiences, in this case, to help teacher candidates see connections between their geometry content course and their future as teachers. Alibegović uses animations created by the Thought Experiments in Mathematics Teaching (ThEMaT) project to represent interactions in secondary mathematics classrooms and that are available in the Lesson*Sketch* platform. The experience we focus on uses *The Midpoint Quadrilateral*. Through interacting with these animations, Alibegović's teacher candidates are able to experience how mathematical content knowledge is relevant to leading mathematical discussions.

The first encounter the teacher candidates have with the Midpoint Quadrilateral Theorem is through their own exploration and experimentation. They are initially asked to construct a quadrilateral inside the Geogebra platform, to construct midpoints of each side, connect the consecutive points with segments, and record all the observations they can make in response to an open-ended question. Shortly thereafter they are asked to prove their conjectures and submit those proofs. Using the discussion forum capability of Lesson*Sketch*, Alibegović's experience asks teacher candidates to collaborate on improving the proofs by providing each other feedback on their proofs.

Once the students have sufficient time to think about the theorem and the proof, they are then shown the animation using Media Show. In the animation, the teacher aims to teach a class about the isosceles trapezoid and its midpoint quadrilateral, the rhombus. The teacher introduces the idea of developing a new theorem by reminding the students of the corresponding theorem for rectangles; she then introduces the definition of the isosceles trapezoid, asks the students to come up with properties of isosceles trapezoids, and finally asks them to make conjectures about the midpoint quadrilateral of an isosceles trapezoid. After watching this animation, teacher candidates are asked questions that will get them thinking about the mathematical decisions that underlie the teachers' actions in this classroom interaction. Specifically, they are asked to consider the teacher's rationale for choosing this particular sequencing of ideas. Here are some example excerpts from teacher candidates' responses. These responses show the teacher candidates engaging in consideration of the unfolding of the mathematical ideas and in justifying the actions of the teacher in terms of this unfolding.

- I think to start with a rectangle is a easy way to show that the midpoints form a rhombus. Students can use the Pythagorean theorem to easily solve for, or see what the length of the rhombus will be, and again easily see that they all end up being the same $a^2 + b^2 = c^2$. Moving next to the trapezoid should let students see that there will still be a relationship between the sides made by connecting the midpoints.
- By starting with a rectangle, the students already have some intuition about what is going on when the trapezoid is experimented with. The difference between a rectangle and an isosceles trapezoid is some extension on one side in the x direction. From Proof 2, it is clear that this will keep the inscribed shape a rhombus as it was under the rectangle. This is a sequence that will help the students start to make this connection.

This animation allows these teacher candidates to see and evaluate one particular sequencing of this material, pushing them to consider the role of the mathematics in the interaction. Teacher candidates were then offered an opportunity to develop their own sequencing of the material. In their responses to this prompt, many were influenced by the sequencing outlined in the animation, but conveyed that they thought the scenario was insufficiently student-centered and outlined activities to remedy that.

Alibegović's experience continues with another excerpt from the animation in which students in the same class share their conjectures. Two students make arguments that the midpoint quadrilateral for an isosceles trapezoid is a rhombus and a kite, respectively. The student who presented an informal justification for the kite, Beta, is then told that in fact the answer is a rhombus. Teacher candidates demonstrated dissatisfaction with this development, most notably through their discussion of the role of definition. In focusing on the role of the definition in the classroom interaction, they are giving evidence that the experience helps them articulate connections between the mathematical content of the course and ways in which they as teachers might structure class discussions of such content. Here are three illustrative teacher candidate responses:

- Beta, the student who said the shape is a kite, is wrong because she forgot about the part of the definition that "no opposite sides are congruent," thus why a rhombus cannot be a kite.
- Well, using my definitions, or the definitions in the Geometry text I used, it is a kite and a rhombus. So, I guess I would have to know what definitions we are using. I would not have told Beta she was wrong, especially because she is using good reasoning on why it is a kite.
- If I were the teacher I would have asked Alpha and Beta to give me their definition of a kite and a rhombus. My definition of a kite is a quadrilateral whose adjacent sides are congruent. The textbook I was using said a kite was a

quadrilateral whose adjacent sides are congruent and whose opposing sides are not congruent. If we were to go with my definition, then a rhombus would be a specific type of kite, but if we were to go off the textbook's definition a rhombus would not be a kite. [...] This is definitely something the class would need to discuss and not just glaze over.

By responding to this animation, the teacher candidates are engaging both with the mathematical content of the definitions of a rhombus and a kite, but also with the teaching practices of interpreting and responding to student thinking. Incorporating this representation of teaching into her geometry course allows Alibegović to make connections between mathematical content and the decisions of teaching around lesson planning, supporting mathematical practices of proving and making conjectures, and interpreting and responding to student thinking. Having teacher candidates respond to the prompts about the animation both individually and to one another in forums provides Alibegović with rich assessment data about their understandings of the mathematical content in the context of actual teaching practice.

12.5 Describing the Fellows' Teacher Education Materials

Having described the functionalities that Lesson*Sketch* provides to teacher educators to do their work and having illustrated what two experiences are like and how students interact with them, we turn now to the central question of the chapter: How might one characterize the content of the materials that the LR+D fellows provide to teacher candidates through the experiences in Lesson*Sketch*? This section explores the utility of a set of concepts offered by Grossman et al. (2009) that they describe as pedagogies of practice. They suggest that across fields, professional education can be usefully described as involving *representations* of practice, *decomposition* of the complexity of practice into practices, and then *approximations* of practice. In working with these concepts to describe the curriculum of teacher education, we suggest that these concepts use the word *practice* in three ways (Lampert, 2010, suggests four views of practice, one of which we do not include here). The complexity of teaching practice (singular) is decomposed into the practices (plural) that constitute the curriculum of teacher education methods courses (Fig. 12.1). Then, in order to have teacher candidates develop skills and expertise related to these practices, teacher educators create opportunities for practicing these practices in settings that approximate the complexity of teaching practice (Fig. 12.1).

As the settings for approximating practice become more complex and more closely approximate practice, teacher candidates gradually learn to stitch the practices together into the practice of a well-prepared beginner (Darling-Hammond, 2012).

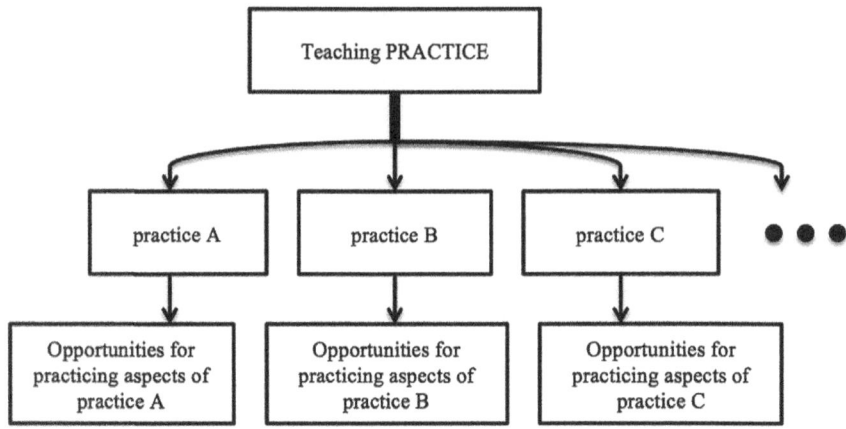

Fig. 12.1 Three uses of practice in practice-based teacher education

In the rest of this section, we explore relationships between Grossman et al.'s (2009) pedagogies of practice and the teacher education curricular artifacts created by the LR+D fellows. With these artifacts, teacher candidates are given a variety of opportunities for practicing aspects of the specific practices that their instructors have chosen as focal. In addition, teacher educators and their students make use of tools provided by the platform to represent the complexity of teaching practice, to represent particular component practices, and to represent candidates' attempts to enact these practices. We begin with how teacher educators represent complexities of practice, move to how they decompose practice into component practices and represent those practices, and then finally to how they create opportunities for teacher candidates to practice skills relevant to these practices and to represent their efforts.

12.5.1 Representing Practice

The use of artifacts of practice in teacher education for representing practice is well established; "Using real artifacts, records, moments, events and tasks permits a kind of study and analysis that is impossible in the abstract" (Ball & Cohen, 1999, p. 24). In this effort, teacher educators collect, organize, and interpret aspects of classroom interaction to represent practice (Herbst et al., 2016, for example, distinguish between found and interpreted representations of practice, as well as ones designed by teacher educators). The artifacts that are created by this activity, like video recordings and written transcripts of classroom dialogue, both make certain aspects of teaching visible to teacher candidates, as well as give teacher candidates an overall sense of the work. Teacher educators also design artifacts—such as cases (Lacey & Merseth, 1993) or animations (Chazan & Herbst, 2012)—to act as representations of practice. The Lesson*Sketch* platform supports work with a wide range of artifacts that represent practice in different ways. In addition to being able to display and use those representations in the context of experiences, the Lesson*Sketch* platform also supports the design and creation of depictions, or storyboards, that are used to represent classrooms with cartoon characters. Below we describe how the Fellows are incorporating representations of practice in their curricular materials.

We begin with ways in which the LR+D Fellows represent instances of practice without decomposing the practice into practices. Subsequent sections will also examine how the fellows represent practice in the context of decomposing practice into practices, as well as how they support their teacher candidates in representing practicing the candidates do. For example, in the context of a course focused on research on learning and its relationship to teaching, in order to make a point about ways in which societal dynamics make themselves felt in the classroom, Fellow Lawrence Clark has designed and created a depiction to invite teacher candidates to interact with a story from his own teaching (based on The Case of Mya in Chazan, Herbst & Clark, 2016). In order to meet his goals, in addition to representing classroom interaction, this depiction also includes the teacher's thinking as the story plays out and the dilemmas that arose for him when what seemed initially to be a sensible solution had unanticipated negative consequences. In creating the depiction, Clark represented the teacher and student thoughts (note the thought bubbles in Fig. 12.2) involving race that were then taken up in different ways by teacher candidates as they interacted with the finalized depiction (Herbst et al., 2017). But, in contrast to the work that will be highlighted in future sections, there is not a particular aspect of teaching for teacher candidates to practice. Instead, the focus here is on teaching as a complex and intertwined practice nested within social structures that created dilemmas for teachers that must be managed, but cannot be solved (Lampert, 1985).

As described earlier, Fellow Emina Alibegović, who teaches mathematics content courses, also uses designed representations. To help teacher candidates see connections between a geometry course and their future as teachers, she uses two animations created by the Thought Experiments in Mathematics Teaching (ThEMaT) project and available in the Lesson*Sketch* platform, *The Midpoint Quadrilateral* and *Postulates and Theorems on Parallel Lines*.[2] Similarly, Fellow Orly Buchbinder incorporates high school students' conjectures when doing geometry tasks into her curricular materials for a geometry content course. Without decomposing practice and providing opportunities to practice specific skills in teaching, she uses the student conjectures and the question of whether these conjectures are a coincidence or a representative case of a general rule (Buchbinder, Zodik, Ron, & Cook, 2017) as a way to suggest that teachers' mathematical understandings are a resource for responding to student work.

---

[2]These can be seen among the animations in Lesson*Sketch*'s Original Collection

Fig. 12.2 Depiction showing Mya switching into a predominantly white class. (Graphics are ©
The Regents of the University of Michigan, used with permission.)

Fellows have also used artifacts of practice like video of actual classrooms.
These videos are often purposefully chosen by a teacher educator and perhaps even
carefully edited by a production team. Videos may have some advantages in suggesting the viability of a vision for practice in real classrooms (Brophy, 2004, p. xi).
Fellow Kristen Bieda uses video records of teaching of The Pool Border Problem
(Boaler & Humphreys, 2005) to provide teacher candidates with an opportunity to
see real students engaging in *Constructing viable arguments and critiquing the
reasoning of others* (CCSSO, 2010, Standard for Mathematical Practice Three). In
choosing to use video, and this video in particular, Bieda is representing a particular
kind of reform-minded mathematics teaching practice her teacher candidates may
not otherwise encounter and suggesting that such teaching practice be emulated.

12.5.2   Decomposing Practice into Practices
and Representing Practices

As they strive to develop materials for practice-based teacher education, several of
the fellows have been decomposing teaching practice into constituent practices and
using representations of classroom interaction to engage teacher candidates with

Fig. 12.3 Depiction showing variety of potential student comments during a launch. (Graphics are © The Regents of the University of Michigan, used with permission.)

specific practices. For example, as described earlier, Fellow Janet Walkoe suggests to teacher candidates that attending to and responding to student thinking is an important component of the work of teaching in mathematics classrooms. Similarly, Fellow Rob Wieman focuses on the launching of an exploratory mathematical task as a particular task of teaching (Wieman & Jansen, 2016) that may be overwhelming to a novice teacher. To do so, he begins his module with a representation of the range of initial reactions students may have when they first encounter an exploratory task (see Fig. 12.3).

But, not only has Wieman decomposed teaching into practices of such a grain size as launching a task, he also has developed templates for his teacher candidates to use to assess and then respond to student questions in the context of launching an exploratory task (For a presentation of the rationale for this structure, see http://resourcecenters2015.videohall.com/presentations/537). Wieman uses the closed-ended question tool in the Lesson*Sketch* platform to provide teacher candidates with four options for classifying student questions and six options for potential responses to student questions. The provision of these options is intended to help teacher candidates develop ways of describing a limited set of potential responses that might serve as a resource in their initial forays in launching exploratory tasks. Wieman's decomposition of teaching practice relies on a representation of one of those practices (launching a task) and the platform allows him to engage his students in the decomposition of this practice into discursive moves.

12.5.3   Approximations of Practice and Representing One's Practicing

"Approximations of practice refer to opportunities for novices to engage in practices that are more or less proximal to the practices of a profession" (Grossman et al., 2009, p. 2049). As described in the three subsections below, in the work of the fellows, these opportunities have ranged from asking teacher candidates to choose a response from a specified set of options, to describing how they would respond to fictional students, to carrying out a particular task of teaching in a field placement and creating a representation of what happened.

12.5.3.1   Selecting a Response from a Fixed Set of Options

A number of fellows have created experiences where teacher candidates practice making teaching decisions. Fellow Woong Lim begins one experience with a middle school student's response to an algebraic task. He then provides three different ways in which teacher candidates from earlier years chose to provide feedback and asks his teacher candidates to describe how they would respond as teachers. Similarly, Fellow Karl Kosko has teacher candidates choose between at least two possible depicted teacher actions and then see how the interaction continues. For example, in one module for elementary teacher candidates, teacher candidates are asked to choose between the following two moves: "The teacher needs to encourage additional student participation to have other strategies for how the problem was solved and compare those strategies" or "The teacher needs to press Jasmine to explain her thinking more clearly until Jasmine recognizes something is not working with her explanation" (Kosko, 2016, p. 1342). The response they select determines what they will see next in the depicted classroom scenario. This type of 'choose your own adventure' activity allows teacher candidates to play out different scenarios to see how different teacher decisions can influence the course of a lesson, without the high stakes of interactions with actual students (Kosko, 2014).

12.5.3.2   Respond as a Teacher

Many of the fellows ask their teacher candidates to respond to questions about some aspect of classroom interaction from the perspective of a teacher. For example, coming from a complex instruction perspective (Cohen & Lotan, 2014), Fellow Sandra Crespo has created an experience designed to help teacher candidates recognize students' mathematical strengths (Featherstone et al., 2011) and to practice articulating those strengths (see Bannister, Kalinec-Craig, Bowen, & Crespo, 2018). Teacher candidates are asked to fill in the following sentence frame to identify potential strengths in the students' mathematical thinking from a given depiction: *It was smart when <name of student> did/said <evidence from the depiction>, because it <how does strength support students' math learning>*. By completing such sentence frames, teacher candidates are given an opportunity to practice responding to student thinking, but with a particular goal and structure in mind. She aims to have teacher candidates develop awareness of deficit frames and language they may bring to describing students' thinking and to provide them with support to begin the process of shifting those frames and language.

Another Fellow, William Zahner, also asks his teacher candidates to respond to a depiction as if they were the teacher. In Zahner's case, he asks teacher candidates to focus on a specific mathematical concept—defining integer exponents. Zahner creates a depiction where two students provide explanations for different ways they are thinking about a^0 (Fig. 12.4).

Fig. 12.4 1 or 0? Articulating that students have different ideas. (Graphics are © The Regents of the University of Michigan, used with permission.)

Fig. 12.5 Teacher candidates asked to respond to a situation where two students disagree. (Graphics are © The Regents of the University of Michigan, used with permission.)

Then, as shown in Fig. 12.5, Zahner asks the teacher candidates to assume they are the teacher and to explain what they would do next to respond to these ideas.

12.5.4 *"Before" and "After" Depictions*

Finally, not only can teacher educators ask their students to respond to particular depictions of practice, but they can also ask them to create storyboards as a way both to indicate what they might do when facing such a situation and what they actually did when facing such a situation in a field placement. Fellow Joel Amidon has been thinking about the Lesson*Sketch* depictions in his module as coming from a "virtual field placement" that his teacher candidates visit three times over the course of the semester (Amidon & Casey, 2016; Amidon, Chazan, Grosser-Clarkson, & Fleming, 2017). In the context of work on eliciting student thinking, he assigns a common interview task to be carried out by each teacher candidate in their own field placement. He then asks his teacher candidates to depict both how they expect their interaction with their interviewee to go and later how it actually transpired.

Similarly, another Fellow, Wendy Rose Aaron, has her students make story-boards depicting their predictions about how enacting a high-leverage instructional practice (e.g., eliciting and responding to student contributions) will go in their classroom (Herbst et al., 2016). Aaron's students are in-service teachers, so she asks them to enact the lesson and to video record the enactment. The teachers are then asked to choose an excerpt from the lesson where they are enacting the focal practice and to create a cartoon depicting what actually happened. The teachers post their depictions to a class forum, where they are asked to comment on the depictions created by their peers.

Both Amidon and Aaron suggest that to have teacher candidates learn from making these depictions it is important for teacher candidates to compare their "before" and "after" depictions and reflect on these differences.

12.6 Summary and Concluding Thoughts

To summarize, in the production of their materials for practice-based, online teacher education, the Lesson*Sketch* Research + Development Fellows have decomposed teaching practice into core practices of teaching and have created opportunities for the practicing of component skills of these core practices. The fellows have had teacher candidates practice by selecting choices from a fixed set of options, by describing how they would respond as teachers, and by asking teacher candidates to carry out a practice with students and then represent what happened. Throughout this work, fellows have engaged in, and have asked their teacher candidates to engage in, representing practice. The representations of practice that they have created represent the complexity of teaching practice, core practices of teaching, and the attempts of teacher candidates to practice component skills of these core practices. As the fellows have done so, the capacities of the Depict tool in Lesson*Sketch* to design representations have been useful to represent practice in all of these ways. As a result of the fellows' work in this digital environment, the pedagogies of practice described by Grossman and her colleagues have come crystallized into curricular artifacts: representing practice has led to the creation of representations of practice, decomposing practice has led to representations of decompositions of practice, and approximating practice has led to representations of opportunities for practicing teaching. In this sense, the concepts that Grossman et al. (2009) described as pedagogies of practice have turned out to be useful in describing the curricular artifacts created by the LR+D fellows.

Acknowledgements This publication was supported by the US National Science Foundation Grant DRL-1316241. The views presented are those of the authors and do not necessarily represent the views of the Foundation.

We greatly appreciate the work of the LR+D fellows and that they have allowed us to represent their work.

The storyboards presented here were created with the Lesson*Sketch* platform. Lesson*Sketch* is designed and developed by Pat Herbst, Dan Chazan, and Vu-Minh Chieu with the GRIP lab, School of Education, University of Michigan. The development of this environment has been supported with funds from National Science Foundation grants ESI-0353285, DRL-0918425, DRL-1316241, and DRL-1420102. The graphics used in the creation of these storyboards are © 2015 The Regents of the University of Michigan, all rights reserved. Used with permission.

References

Amidon, J., & Casey, S. (2016, January). *Developing preservice teacher noticing via the Lesson*Sketch *platform*. Paper presented at the meeting of the Association of Mathematics Teacher Educators, Irvine, CA.

Amidon, J., Chazan, D., Grosser-Clarkson, D., & Fleming, E. (2017). Meet me in Azul's room: Designing a virtual field placement for learning to teach mathematics. *Mathematics Teacher Educator, 6*(1), 52–66.

Ball, D. L., & Cohen, D. K. (1999). Developing practice, developing practitioners: Toward a practice-based theory of professional education. In L. Darling-Hammond & G. Sykes (Eds.), *Teaching as a learning profession: Handbook of policy and practice* (pp. 3–32). San Francisco: Jossey-Bass.

Ball, D. L., & Forzani, F. M. (2009). The work of teaching and the challenge for teacher education. *Journal of Teacher Education, 60*(5), 497–511. https://doi.org/10.1177/0022487109348479.

Ball, D. L., Sleep, L., Boerst, T. A., & Bass, H. (2009). Combining the development of practice and the practice of development in teacher education. *The Elementary School Journal, 109*(5), 458–474. https://doi.org/10.1086/596996.

Bannister, N., Kalinec-Craig, C., Bowen, D. & Crespo, S. (2018). Learning to notice and name students' mathematical strengths: A digital experience. *Journal of Technology and Teacher Education, 26*(1), 13–31. Waynesville, NC USA: Society for Information Technology & Teacher Education.

Boaler, J., & Humphreys, C. (2005). *Connecting mathematical ideas: Middle school video cases to support teaching and learning*. Portsmouth, NH: Heinemann.

Brophy, J. (Ed.). (2004). *Using video in teacher education*. Oxford: Elsevier.

Buchbinder, O., Zodik, I., Ron, G., & Cook, A. (2017). What can you infer from this example? Applications of online, rich-media tasks for enhancing pre-service teachers' knowledge of the roles of examples in proving. In A. Leung & A. Baccaglini-Frank (Eds.), *Digital technologies in designing mathematics education tasks: Potential and pitfalls* (pp. 215–235). Switzerland: Springer. https://doi.org/10.1007/978-3-319-43423-0_11.

Chazan, D., & Herbst, P. (2012). Animations of classroom interaction: Expanding the boundaries of video records of practice. *Teachers' College Record, 114*(3), 1–34.

Chazan, D., Herbst, P., & Clark, L. (2016). Research on the teaching of mathematics: A call to theorize the role of society and schooling in mathematics instruction. In D. Gitomer & C. Bell (Eds.), *Handbook of research on teaching* (5th ed., pp. 1039–1098). Washington, DC: American Educational Research Association.

Chazan, D., & Yerushalmy, M. (2014). The future of textbooks: Ramifications of technological change for curricular research in mathematics education. In M. Stochetti (Ed.), *Media and education in the digital age: A critical introduction* (pp. 63–76). New York, NY: Peter Lang.

Cohen, E. G., & Lotan, R. A. (2014). *Designing groupwork: Strategies for the heterogeneous classroom* (3rd ed.). New York, NY: Teachers College Press.

Darling-Hammond, L. (2012). The right start: Creating a strong foundation for the teaching career. *Phi Delta Kappan, 94*(3), 8–13. https://doi.org/10.1177/003172171209400303.

Featherstone, H., Crespo, S., Jilk, L. M., Oslund, J. A., Parks, A. N., & Wood, M. B. (2011). *Smarter together! Collaboration and equity in the elementary math classroom* (1st ed.). Reston, VA: National Council of Teachers of Mathematics.

Grossman, P., Compton, C., Igra, D., Ronfeldt, M., Shahan, E., & Williamson, P. (2009). Teaching practice: A cross-professional perspective. *Teachers College Record, 111*(9), 2055–2100.

Grossman, P., & McDonald, M. (2008). Back to the future: Directions for research in teaching and teacher education. *American Educational Research Journal, 45*(1), 184–205.

Herbst, P., Aaron, W., & Chieu, V. M. (2013). Lesson*Sketch*: An environment for teachers to examine mathematical practice and learn about its standards. In D. Polly (Ed.), *Common core mathematics standards and implementing digital technologies* (pp. 281–294). Hershey, PA: IGI Global.

Herbst, P., Boileau, N., Clark, L., Milewski, A., Chieu, V. M., Gursel, U., et al. (2017). *Directing focus and enabling inquiry with representations of practice: Written cases, storyboards, and teacher education.* Paper accepted and to be presented at 39th annual meeting of the North American Chapter of the International Group for the Psychology of Mathematics Education. Indianapolis, IN.

Herbst, P., Chazan, D., Chieu, V. M., Milewski, A., Kosko, K. W., & Aaron, W. R. (2016). Technology-mediated mathematics teacher development: Research on digital pedagogies of practice. In M. Niess, S. Driskell, & K. Hollebrands (Eds.), *Handbook of research on transforming mathematics teacher education in the digital age* (pp. 78–106). Hershey, PA: IGI Global. https://doi.org/10.4018/978-1-5225-0120-6.ch004.

Herbst, P., & Chieu, V. M. (2011). *Depict: A tool to represent classroom scenarios.* Technical report. Deep Blue at the University of Michigan. http://hdl.handle.net/2027.42/87949.

Kosko, K. W. (2014). Using multi-decision scenarios to facilitate teacher knowledge for mathematical questioning. In M. J. Mohr-Schroeder & S. S. Harkness (Eds.), *Proceedings of the 113th Annual Convention of the School Science and Mathematics Association* (pp. 23–30). Jacksonville, FL: SSMA.

Kosko, K. W. (2016). Improving pre-service teachers' noticing while learning to launch. In M. B. Wood, E. E. Turner, M. Civil, & J. A. Eli (Eds.), *Proceedings of the 38th Annual Meeting of the North American Chapter of the International Group for the Psychology of Mathematics Education* (pp. 1341–1344). Tucson, AZ: The University of Arizona.

Lacey, C. A., & Merseth, K. K. (1993). Cases, hypermedia and computer networks: Three curricular innovations for teacher education. *Journal of Curriculum Studies, 25,* 543–551.

Lampert, M. (1985). How do teachers manage to teach? Perspectives on problems in practice. *Harvard Educational Review, 55*(2), 178–194.

Lampert, M. (2010). Learning teaching in, from, and for practice: What do we mean? *Journal of Teacher Education, 61*(1–2), 21–34. https://doi.org/10.1177/0022487109347321.

Love, E., & Pimm, D. (1996). 'This is so': A text on texts. In A. Bishop, K. Clements, C. Keitel, J. Kilpatrick, & C. Laborde (Eds.), *International handbook of mathematics education* (Vol. 1, pp. 371–409). Dordrecht, The Netherlands: Kluwer.

McDonald, M., Kazemi, E., & Kavanagh, S. S. (2013). Core practices and pedagogies of teacher education: A call for a common language and collective activity. *Journal of Teacher Education, 64*(4), 378–386.

National Governors Association Center for Best Practices & Council of Chief State School Officers. (2010). *Common core state standards for mathematics.* Washington, DC: Authors.

Popkewitz, T. S. (1987). *The formation of school subjects: The struggle for creating an American institution.* London, UK: Falmer Press.

Sherin, M., & van Es, E. (2009). Effects of video club participation on teachers' professional vision. *Journal of Teacher Education, 60*(1), 20–37.

Walkoe, J. & Levin, D.M. (2018). Using technology in representing practice to support preservice teachers' quality questioning: The roles of noticing in improving practice. *Journal of Technology and Teacher Education, 26*(1), 127–147. Waynesville, NC USA: Society for Information Technology & Teacher Education.

Wieman, R., & Jansen, A. (2016). Improving pre-service teachers' noticing while learning to launch. In M. B. Wood, E. E. Turner, M. Civil, & J. A. Eli (Eds.), *Proceedings of the 38th Annual Meeting of the North American Chapter of the International Group for the Psychology of Mathematics Education* (pp. 837–844). Tucson, AZ: The University of Arizona.

Yerushalmy, M. (2005). Functions of interactive visual representations in interactive mathematical textbooks. *International Journal of Computers for Mathematical Learning, 10*(3), 217–249. https://doi.org/10.1007/s10758-005-0538-2.

Zeichner, K. (2012). The turn once again toward practice-based teacher education. *Journal of Teacher Education, 63*(5), 376–382. https://doi.org/10.1177/0022487112445789.

The manufacturer's authorised representative in the EU is Springer
Nature Customer Service Centre GmbH, Europaplatz 3, 69115 Heidelberg,
Germany. If you have any concerns regarding our products, please
contact ProductSafety@springernature.com

Printed and bound by CPI Group (UK) Ltd, Croydon, CR0 4YY
23/04/2026
02095592-0011